FLOWERS

First published in 2013 by New Holland Publishers

London • Cape Town • Sydney • Auckland

www.newhollandpublishers.com

Garfield House, 86–88 Edgware Road, London W2 2EA, UK

Wembley Square First Floor, Solan Road Gardens, Cape Town 8001, South Africa

Unit 1, 66 Gibbes Street, Chatswood, NSW 2067, Australia

218 Lake Road, Northcote, Auckland, New Zealand

10 9 8 7 6 5 4 3 2 1

A CIP catalogue record for this book is available from the British Library.

ISBN 978 1 78009 261 4

Senior Editors: Krystyna Mayer, Sally McFall
Designer: Kimberley Pearce
Production: Olga Dementiev
Publisher: Simon Papps

Printed and bound in China by Toppan Leefung Printing Ltd

POCKET GUIDE TO

WILD
FLOWERS

BOB GIBBONS

NEW
HOLLAND

CONTENTS

Introduction

This book is intended as a handy guide that can readily be taken out into the countryside and used to identify some of the most attractive, abundant and conspicuous flowers of Britain and adjacent parts of Europe. It does not, of course, cover all the flowers of this area, and the selected species are ones that you are generally most likely to see. Where appropriate, a selection of similar species is included with the main species.

The order of species within the book follows the most widely used taxonomic system. At present, botanical naming is in a state of considerable flux because of the vast amounts of new information arising from DNA analysis of plants. The process is by no means complete but has already led to much reclassifying and renaming of species. This book follows the nomenclature of the *New Flora of the British Isles*, third edition, by C. Stace (see **Bibliography and Resources**, p. 189); older names are also given where they are still widely used.

How to use this book

The species are described in a standardized way, comprising a simple description of the plant, its flowering time, usual habitat preferences and broad distribution within the British Isles including Eire. What is included in the description varies – the salient points that distinguish each species are covered. For example, in the mulleins there is a description of the colour of the hairs on the filaments of the stamens because this feature is important for identifying mulleins, but this characteristic is not described elsewhere. The flowering time indicated is the main period of flowering, but plants are likely to flower latest further north and at high altitudes, and flowering periods vary from year to year. In addition, small numbers of flowers may be found well outside the normal period of flowering, so the dates given should be treated as a guide only. The habitat preferences given are the most frequent ones and may not cover all the situations in which a plant occurs, and plants may grow in different habitats in different countries.

Following the description of the main species on the page, a number of similar or closely related species may be described. The descriptions are short, simply outlining the key differences from the main species.

The book is intended primarily as an accessible visual guide, with the photographs giving the primary clue to each flower's identity, to be confirmed by study of the more detailed descriptions and similar species where applicable.

Foxglove flowers *Digitalis purpurea*

The structure of flowers

Flowers vary hugely in their form and colour, depending on how they are pollinated, where they grow and how they have evolved. It is not possible here to describe and explain the whole range of flower structures, but there are some simple, general rules that can be followed.

Flowers consist of several whorls of parts that may or may not be symmetrically arranged. The outermost whorl (which is often, but not necessarily, green) is known as the **calyx**. This is made up of individual **sepals** that may be separate or fused into a tube with just the tips free, indicating the number of sepals making up the tube. The primary function of the calyx is to protect the bud, though in many species the sepals are adapted to be part of the pollination mechanism. In some species the calyx may be missing or adapted to be similar to the petals – for example in anemones, where there is no obvious calyx, and in many orchids, where the sepals are petal-like and are significant parts of the insect-attracting mechanism. In a few plant groups, notably the rose and mallow families, there is an additional epicalyx outside the calyx, but this is the exception rather than the rule. Its presence can be helpful in identifying these families.

The next whorl in towards the centre of a flower is composed of the **petals**, which are most commonly what we think of as the flower. They are frequently highly coloured and conspicuous, and form the key feature in attracting insects to pollinate. The form of petals, and their arrangement, varies enormously. One common form is a simple ring of roughly equal petals as, for example, in cranesbills and buttercups; in other species, such as the bellflowers, the petals are fused into a tube. In many groups, the petals have evolved individually and are no longer all similar in shape – in the pea family, one petal forms the large raised standard and two form projecting wings, enclosing two more fused into a keel. An extreme variation occurs in the orchids, especially in the genus *Ophrys*, such as the Bee Orchid on p. 181. Here the three sepals are usually strongly coloured and resemble petals; the lowest petal, known as the labellum or lip, has developed into a form resembling the body of a bumblebee, while the remaining two petals are small, looking something like the antennae of an insect. The whole is part of a complex process of duping male bees and wasps into thinking that the plant is a female insect ready for mating – a process known as pseudocopulation.

Within the ring of petals lie the reproductive parts of the flower. The male parts are the **stamens**, usually each consisting of an **anther**, where the pollen is produced, and a **filament** – that is, its stalk. The male genes are carried in the pollen by wind, insects or other means. The female parts are more variable, made up of one to many **carpels**, each containing one or more **ovaries**. A carpel normally has a stalk-like **style** that terminates in a receptive **stigma**, where the pollen settles and germinates. The number and arrangement of these parts is an important component of plant identification.

The leaves of plants are also important in their identification, particularly their shape, whether they are toothed or not and whether stalked or not, and whether they are arranged in opposite pairs, alternately up the stem, or in some other way. The presence, absence and shape of two little leaf-like structures at the base of the leaf stalk – known as **stipules** – are also important.

Wild plants are under threat everywhere, and many have declined alarmingly. Please pick them only if necessary, never dig them up and join as many wildlife conservation organizations as you can.

The distinctive flowers of Bluebell *Hyacinthoides non-scripta*

WHITE WATER-LILY
Nymphaea alba

One of northern Europe's most distinctive and attractive wild plants, White Water-lily is an aquatic perennial with large, leathery, almost circular floating leaves that have veins radiating from the centre, where the stalk joins the blade. The flowers are very large, up to 20cm across, scented, with numerous white petals, some of which are longer than the four green sepals. They open fully only in bright conditions. The fruits are ovoid to spherical and fleshy, though rarely seen because they ripen below the water's surface. When mature, the fruits sink to the bottom and decay to release the seeds.

FLOWERING TIME June–August.

DISTRIBUTION Widespread and common in lakes, ponds and canals throughout Britain and Ireland wherever there is suitable habitat.

SIMILAR SPECIES **Yellow Water-lily *Nuphar lutea*** is easily distinguished when in flower by its smaller yellow flowers, with both the flowers and fruits held well above water level. The leaves differ in being more oval, with a herringbone pattern of veins. The species is widespread in similar habitats throughout the region, except uplands.

White Water-lily *Nymphaea alba*

MARSH MARIGOLD
Caltha palustris

A familiar spring flower of damp and marshy places, Marsh Marigold, or Kingcups, is a robust hairless perennial up to 50cm tall, with thick, hollow stems and shiny, kidney-shaped leaves, of which the lower are stalked, the upper unstalked. The flowers are large, 2–5cm across, and bright golden-yellow. They are made up of five or more petal-like sepals (if you look underneath the flower, there is no obvious green calyx, which helps to distinguish this plant from buttercups, *see* p. 18). The fruits form a cluster of distinctive pod-like, multi-seeded, erect follicles. Like buttercups, the plants are poisonous, and the flowers have occasionally been used to produce a yellow dye. Marsh Marigold is frequently grown in gardens as an attractive ornamental plant, often in double-flowered forms.

FLOWERING TIME March–July.

DISTRIBUTION Widespread and frequently abundant in a variety of wet places, both shaded and sunny, including marshes, fens, wet woodland and pastures, from sea level up to at least 1,000m in mountain areas. Generally more abundant in western areas, where suitable habitats are more frequent.

Marsh Marigold *Caltha palustris*

WOOD ANEMONE
Anemone nemorosa

Wood Anemones are well known and much loved as one of the first and prettiest flowers of spring, often seen in great abundance in woodland before the leaves of the trees emerge. They are low-growing, usually hairless perennials, rarely taller than 25cm, spreading by rhizomes to form dense clumps. The flowers are white, though often tinged with pink or purple, and about 2–4cm across. As in Marsh Marigold (and some other plants in the buttercup family), the five or more 'petals' are actually sepals, with no green calyx below them. The leaves are deeply divided into three or more lobes, sometimes appearing after the flowers, then disappearing quite soon after flowering.

FLOWERING TIME March–May.

DISTRIBUTION
A common plant of woodland, particularly ancient woodland, Wood Anemone flowers spectacularly in early spring. It may also appear in non-woodland habitats such as pastures, hedge-banks and cliff-top grassland, often indicating the past presence of woodland. It is widely cultivated in gardens, especially in the strongly coloured pink and purplish varieties.

Wood Anemone *Anemone nemorosa*

STINKING HELLEBORE
Helleborus foetidus

This strong-growing, evergreen perennial up to 80cm tall has an unpleasant smell, as its name suggests. There are no basal leaves, but the stem leaves are palmately divided with simple, narrow, dark green, toothed lobes. The flowers are yellowish-green with a red or purple rim, nodding, bell-shaped and 1–3cm across, usually growing in clusters. The bracts that support the flowers are simple and undivided. A cluster of three inflated, beaked fruits, or follicles, develops in the centre after flowering.

FLOWERING TIME January–April.

DISTRIBUTION An uncommon plant of woodland and scrub, mainly on calcareous soil. Local and mainly in western Britain (though also common as a garden escape); absent from Ireland as a native.

SIMILAR SPECIES Green Hellebore *H. viridis* differs in having basal leaves, deeply divided bracts and more open, greener flowers. It is an uncommon native in western Britain.

Stinking Hellebore *Helleborus foetidus*

TRAVELLER'S JOY
Clematis vitalba

Traveller's Joy is a strong-growing, robust, deciduous, perennial woody climber that in favourable conditions can achieve heights of around 30m, reaching to the top of whatever supports it. The leaves are pinnate, with a few toothed leaflets and twining stems and stalks. The flowers are greenish-white, fragrant and 1–2cm across, grouped into large, dense inflorescences. The four petal-like structures are actually sepals. The plant is most distinctive after flowering, when the fruits develop long, silky plumes (hence another common name for the plant, Old Man's Beard), which remain conspicuous from late summer into winter. Collectively, the plants are highly visible and distinctive.

FLOWERING TIME July–August.

DISTRIBUTION Widespread and common in lowland calcareous places, where it is sometimes considered a weed.

Traveller's Joy *Clematis vitalba*

CREEPING BUTTERCUP
Ranunculus repens

Creeping Buttercup is an all-too-familiar plant of gardens and almost any other habitat, where it forms large, dense patches due to its creeping runners that spread and root. It is a hairy perennial up to about 50cm tall, with triangular leaves each divided into three lobes, of which the central one is stalked. The flowers are golden-yellow, on furrowed stalks, with a variable number of petals, usually between five and seven; the sepals are erect and pressed against the petals.

FLOWERING TIME May–August.

DISTRIBUTION Common as a garden weed throughout Britain and Ireland, it can also be found in grassland, open woodland and arable fields, most frequently where the soil is damp or poorly drained.

SIMILAR SPECIES There are many other similar species of buttercup, broadly the same in appearance but differing in details. One of the most common, **Meadow Buttercup *R. acris***, is more erect, not creeping, with the central lobe of the leaf unstalked and with unfurrowed flower stalks. Another common species, **Bulbous Buttercup *R. bulbosus***, likes drier situations. It has a bulbous base to the stem and the flower sepals are strongly reflexed back down the stem. All three species thrive in open, grassy habitats.

Creeping Buttercup *Ranunculus repens*

LESSER CELANDINE
Ficaria verna (Ranunculus ficaria)

A widespread, familiar, early-flowering, herbaceous perennial, Lesser Celandine is low-growing, rarely reaching 25cm tall, with small, glossy, heart-shaped, long-stalked leaves and tuberous roots. The flowers are bright, shiny yellow, similar to buttercups but with more petals (7–12) and fewer sepals (three instead of five). Although perennial, the plants die back and disappear soon after flowering.

FLOWERING TIME February–May.

DISTRIBUTION Common to abundant throughout Britain and Ireland in a wide variety of habitats up to 750m, including woodland, grassland, churchyards, gardens and roadsides, most often on relatively heavy, damp soils that are partly shaded.

Lesser Celandine *Ficaria verna (Ranunculus ficaria)*

POND WATER-CROWFOOT
Ranunculus peltatus

The water-crowfoots are distinctive as a group, though it can be hard to identify individual species. Collectively, they are familiar as white-flowered, often floating plants of open water and wet places. Pond Water-crowfoot may have rounded, lobed, floating leaves, or finely divided, thread-like submerged leaves, or both, according to circumstances. The flowers are clear white, solitary, 1.5–2cm across and wide open, with five petals, masses of yellow stamens and small, pear-shaped nectaries. There are many other similar species, differing in small botanical details.

FLOWERING TIME May–August.

DISTRIBUTION Widespread and common throughout most of Britain and Ireland, though less common in the north or in upland areas, in a wide range of wetland habitats including slow-flowing rivers, ponds, lagoons and canals.

Pond Water-crowfoot *Ranunculus peltatus*

COLUMBINE
Aquilegia vulgaris

This is an attractive and distinctive perennial up to 90cm tall, and occasionally taller. The leaves are dull green, twice ternate (three-lobed, with each lobe divided again into three lobes) and long stalked. The flowers are large, conspicuous and blue-purple; they are made up of five petal-like sepals, and five petals each prolonged upwards into a curved, hooked spur with a swollen tip. In the centre, a cluster of yellow stamens protrude just beyond the petals. The fruits are erect and stickily hairy. Where populations of the flower are of garden origin, the incidence of different flower colours – such as pink, red and white – is much more frequent than among wild plants. The flowers are pollinated by long-tongued bees that can reach into the spurs.

FLOWERING TIME May–July.

DISTRIBUTION Widespread and common throughout the area in lowland, woodland and scrub, particularly on calcareous soil, as well as on fens and marshes, though it is probable that the majority of populations derive from garden escapes.

Columbine *Aquilegia vulgaris*

Common Poppy
Papaver rhoeas

Common Poppy is a distinctive and popular annual up to about 60cm tall. The stems are bristly hairy (with the hairs usually projecting at right angles to the stem) and the leaves are coarsely pinnately lobed. The buds are drooping and enclosed by two robust, bristly sepals that soon fall, straightening up as they open to reveal the bold scarlet flowers, which are made up of four petals, each often with a black blotch at its base. In the centre there is a cluster of bluish anthers surrounding the developing ovary. The fruit is single, hairless and has a roughly wine-glass shape.

FLOWERING TIME June–September.

DISTRIBUTION Widespread and common throughout lowland areas of Britain and Ireland in open habitats, especially arable fields, waste ground and roadsides, though probably declining as a result of increased pesticide use and improved seed-cleaning methods. It is now quite commonly grown as a garden plant, forming part of annual mixes.

A classic field of poppies

Common Poppy *Papaver rhoeas*

SIMILAR SPECIES There are several similar species, but the closest is **Long-headed Poppy *P. dubium***, which has (usually) paler, more orange-coloured flowers, a longer, narrower fruit capsule and at least some of the stem hairs pressed against the stem. It has a similar distribution, in the same habitats, but is less common.

GREATER CELANDINE
Chelidonium majus

This slender, erect, branched, almost hairless perennial is up to about 90cm tall, and notable for the bright orange latex that it produces when it is cut or broken. The leaves are pale green and coarsely pinnately lobed, and have rounded lobes or teeth. The flowers are orange-yellow and 2–2.5cm across, with two deciduous sepals and four petals; they are borne in few-flowered umbels. The fruit is a long, narrow, hairless capsule. The plant was originally used for medicinal purposes, including in the removal of warts and for eye infections, though all parts of it are actually poisonous.

FLOWERING TIME May–September.

DISTRIBUTION In Britain this is almost certainly an ancient introduction; it has been recorded since Roman times and is now known from a variety of semi-shaded habitats such as gardens, roadsides and woodland edges. It occurs throughout the lowland areas of Britain, though its distribution in Ireland is more scattered.

SIMILAR SPECIES Welsh Poppy *Meconopsis cambrica* is closely related but has much larger, solitary and slightly paler yellow flowers. It is an uncommon native plant in south-west England, Wales and scattered localities in Ireland, but it is also widely naturalized elsewhere as a garden escape.

Greater Celandine *Chelidonium majus*

Common Fumitory
Fumaria officinalis

The fumitories are a distinctive group of sprawling or scrambling annual plants, with delicate greyish leaves, divided pinnately, and white to pale pink or purplish flowers, most commonly held in a long raceme. The flowers are unusual, with two sepals, four petals with one or two spurs, and two stamens. The fruits are egg-shaped to spherical and about 2mm in diameter. Though distinctive as a group, the individual species need close examination for certain identification. The Common Fumitory has flat leaves, purplish flowers in a dense inflorescence, and notched fruit.

FLOWERING TIME May–October.

DISTRIBUTION It is common, especially in the south, as a weed of sandy and chalky places.

SIMILAR SPECIES
The **Dense-flowered Fumitory *Fumaria densiflora*** has more broadly rounded sepals and starts out with dense racemes that eventually become more widely spaced. The **White Ramping Fumitory *Fumaria capreolata*** differs from the Common Fumitory in that it is climbing, with fruit stalks that turn downward.

Common Fumitory *Fumaria officinalis*

FIELD MOUSE-EAR
Cerastium arvense

This low-growing, variable, downy perennial grows in loose mats with both prostrate and ascending stems. The leaves are narrow and pointed, in opposite pairs. The white flowers are larger than most mouse-ears – up to 2cm across, with five deeply notched petals about twice as long as the sepals – and are held in loosely branched groups, often collectively forming a densely flowery mat.

FLOWERING TIME May–August.

DISTRIBUTION Common in Britain on well-drained soils in grassy places such as sand dunes, roadsides, downland and meadows, though rarer in the west.

SIMILAR SPECIES There are many other species of chickweed, of which all of the lowland species have smaller flowers, usually less than 1cm across.

Field Mouse-ear *Cerastium arvense*

GREATER STITCHWORT
Stellaria holostea

This sprawling, rather fragile perennial has weak, square and toothed stems. The narrow leaves are pointed, in opposite pairs, and have finely toothed margins. The white flowers are starry and flat to cup-shaped; the five petals are each divided, so that a flower can appear to have ten petals. They grow in loose clusters. When growing in dense masses, Greater Stitchwort is a very attractive plant.

FLOWERING TIME April–June.

DISTRIBUTION An abundant, widespread spring-flowering plant, occurring throughout Britain and most of Ireland in woodland, scrub, roadsides and other shady habitats, particularly on slightly acid soils.

SIMILAR SPECIES There are a number of species, but the closest is **Lesser Stitchwort *S. graminea***, which differs in having smaller flowers, smooth stems and leaf margins, and occurs in more open, grassy habitats. It flowers later and continues through the summer. **Common Chickweed *S. media*** is annual, with a single line of hairs down the stem, changing sides at each node, and small flowers with deeply divided petals. **Water Chickweed *Myosoton aquaticum*** is similar but with broader leaves, five styles (the others have three) and stickily hairy flower stalks; it occurs in wet places in southern Britain.

Greater Stitchwort *Stellaria holostea*

RED CAMPION
Silene dioica

Red Campion is an erect, hairy, biennial or herbaceous perennial up to about 90cm tall. The leaves are oblong to oval, stalked at the base of the plant and unstalked higher up, and grow in opposite pairs. The flowers are bright red to rose-pink, 2–2.5cm across and have five petals, each divided about halfway to the middle. The flowers are dioecious – that is, either male or female, usually borne on separate plants; the sexes can be distinguished by the sexual parts, and also by the fact that the male calyx tube has 10 veins, while the female calyx has 20.

FLOWERING TIME May–June mainly, though actually recorded in flower in every month of the year.

DISTRIBUTION Common or abundant throughout Britain, though rare in Ireland, in a variety of habitats including woodland, gardens, hedge-banks and cliffs, particularly where they are nutrient-rich.

Red Campion *Silene dioica*

Female Red Campion flowers *Silene dioica*

SIMILAR SPECIES Ragged Robin *Silene flos-cuculi* is rather similar, but less hairy, with narrower leaves and paler pink flowers that have deeply divided, ragged petals. It is widespread but found mainly in damp, open habitats such as marshes and wet meadows. It flowers in May–July.

WHITE CAMPION
Silene latifolia

This erect, stickily hairy, annual to herbaceous perennial is up to 90cm tall. It is similar in general form to Red Campion, but the flowers are pure white and rather larger – up to 3cm across, with a slightly more inflated calyx.

FLOWERING TIME May–September, but varying widely according to location.

DISTRIBUTION Widespread and common in open habitats such as roadsides, cultivated fields and waste ground in lowland areas, especially on less acid soil. It is least common in Wales and Ireland.

SIMILAR SPECIES There are two common, rather similar species. **Bladder Campion *S. vulgaris*** is a hairless plant with smaller white flowers, less than 2cm across, which have deeply notched, separated petals, an inflated, bladder-like calyx and only three styles (not five). It has similar habitats and distribution. **Sea Campion *S. uniflora*** is a prostrate, mat-forming plant rarely growing above 30cm tall. The flowers resemble those of Bladder Campion, but are larger. It is confined mainly to a variety of coastal habitats, especially cliff-tops and shingle banks, as well as a few mountains, all around Britain and Ireland, often producing spectacular displays in early summer.

Bladder Campion *Silene vulgaris*

White Campion *Silene latifolia*

COMMON BISTORT
Persicaria bistorta

Common Bistort is a hairless, clump-forming perennial with erect stems up to 50cm tall growing from creeping rhizomes. The lower leaves are roughly oblong, with winged stems and a papery, encircling collar – the ochrea – where the petiole joins the main stem. The flowers are small, pink and grouped into a dense, conspicuous terminal spike up to 6cm long.

FLOWERING TIME May–August.

DISTRIBUTION Common and widespread in England, Wales and southern Scotland – though rare elsewhere – in damp habitats, especially grassland, roadsides and open woodland. Many of the populations derive from garden escapes, growing in places that are now difficult to distinguish from native sites.

SIMILAR SPECIES Redshank *P. maculosa* is a less glamorous version of Common Bistort, differing in being annual, having more slender, less colourful flower spikes, and leaves that often show a dark, arrow-shaped blotch. It is widespread throughout the region

A meadow brimming with Common Bistort *Pesicaria bistorta*

Common Bistort *Pesicaria bistorta*

in damp, open habitats such as riversides, pond margins and damp waste ground. **Amphibious Bistort *P. amphibia*** usually occurs as a distinctive floating aquatic, but may also be a terrestrial plant. It has short, wide, dense, egg-shaped inflorescences, and leaves that have a truncate or heart-shaped base. It is common throughout the region in wetland habitats, with the exception of higher mountain areas.

THRIFT
Armeria maritima

One of the region's most distinctive and familiar coastal wild plants, Thrift is a cushion-forming perennial that pushes up a dense mass of stems from a thick, woody rootstock. The leaves are dark green, narrow, barely 2mm wide and up to 10cm long, and normally one-veined. They form a dense mat from which unbranched, erect, leafless stems produce the flowers. These are pink, in dense globular heads up to 2.5cm across, with a brown, papery, encircling sheath just below the head; they are borne on leafless stalks up to 30cm tall. Thrift often occurs in great abundance, producing both a spectacular sight and a strong scent from the fragrant flowers.

FLOWERING TIME April–August, though flowering may begin earlier nowadays.

DISTRIBUTION
Almost exclusively a plant of coastal areas and mountains, along with a few other minor open habitats such as riverside gravel and mineral-rich spoil heaps. It is abundant and often dominant in most unspoiled coastal habitats all around Britain and Ireland, though rather rarer inland. It is also grown as a garden plant, from where it may naturalize.

Thrift *Armeria maritima*

IMPERFORATE ST JOHN'S-WORT
Hypericum maculatum

This erect, hairless herbaceous perennial is up to 60cm tall, with roughly square stems that have four raised lines but are neither sharply square nor winged. The roughly oval leaves are unstalked, in opposite pairs and without (or with very few) translucent, dot-like glands. The bright yellow flowers are open, five-parted and about 2cm across; they are dotted with black glands on both the petals and sepals, and borne in loose terminal inflorescences.

FLOWERING TIME June –August.

DISTRIBUTION Widespread and locally common throughout most of Britain and Ireland (rarest in uplands and eastern England) in damp, grassy habitats.

SIMILAR SPECIES There are a number of other very similar species. **Perforate St John's-wort** *H. perforatum* looks very similar, but has roughly oval stems with only two raised lines, and leaves that are conspicuously dotted with small, translucent glands. It is rather similar in terms of habitats and flowering, but more widespread in the east. **Square-stemmed St John's-wort** *H. tetrapterum* has square stems with wings on the four angles, and smaller flowers up to 1cm across. It is common in marshy places throughout the area except the highlands. **Hairy St John's-wort** *H. hirsutum* has round, unlined stems, and is downy or hairy. It is common on calcareous soils except in Ireland and parts of western Britain.

Imperforate St John's-wort *Hypericum maculatum*

COMMON MALLOW
Malva sylvestris

Common Mallow is an erect or sprawling, slightly hairy biennial or perennial. The leaves are stalked, rounded or heart-shaped, each with three to seven shallow, toothed lobes. The pretty flowers are pink to purple with darker longitudinal veins, up to 2.5–4cm across and borne in few-flowered clusters in the leaf axils. There are five each of the petals and sepals, and outside the sepals there is an additional ring of narrow, calyx-like segments – the epicalyx, which is typical of the mallow family.

FLOWERING TIME June–September.

DISTRIBUTION Probably an ancient introduction to Britain and now widespread and generally common in England, Wales and Ireland, though becoming less frequent northwards and westwards. It occurs in a variety of sunny habitats including meadows, waste-ground road verges and coastal habitats such as shingle beaches.

Common Mallow *Malva sylvestris*

Musk-mallow *Malva moschata*

SIMILAR SPECIES **Musk-mallow *Malva moschata*** differs in having deeply divided leaves and paler flowers. **Tree-mallow *Lavatera arborea*** has flowers similar to Common Mallow's, but borne on a tall, woody, bushy plant, up to 3m tall, and is confined largely to western coastal habitats, though also widely planted and naturalized from gardens. **Marsh-mallow *Althaea officinalis*** is a tall, greyish-hairy plant with larger and paler pink flowers up to 4cm across, confined as a native plant to southern coastal parts of Britain, but also widely planted and naturalized.

ROUND-LEAVED SUNDEW
Drosera rotundifolia

The sundews are a distinctive group of small perennial bog plants notable for their insect-eating activity. Round-leaved Sundew consists of a rosette of leaves shaped roughly like a table-tennis bat, of which the blade part is rarely more than 1cm across, covered with sticky, red glandular hairs that play a key role in catching insects. From the centre of the rosette, an erect, leafless stalk arises, bearing a loose cluster of white flowers, each about 5mm across and with five small petals. The inflorescence rises well above the leaf rosette.

FLOWERING TIME June–August.

DISTRIBUTION Sundews in Britain are confined to wet, acidic habitats, especially on peat, such as bogs, pool margins and wet heaths. Round-leaved Sundew is widespread and generally common in northern and western Britain, and the whole of Ireland, but becomes steadily rarer eastwards.

Oblong-leaved Sundew *Drosera intermedia*

Round-leaved Sundew *Drosera rotundifolia*

SIMILAR SPECIES Two other species occur in Britain and Ireland.
Oblong-leaved Sundew *D. intermedia* has narrower, almost
parallel-sided leaves, and the flowers arise on curled stalks from
below the rosette. It has similar habitats and distribution, though it
is less common. **Great Sundew *D. anglica*** has leaves of the same
shape as the Oblong-leaved Sundew, but up to 3cm long, and the
tall inflorescence arises in the centre of the rosette to produce larger
flowers. It occurs in similar habitats, though usually wetter, and is
virtually restricted to northern and western Britain and Ireland.

COMMON ROCK-ROSE
Helianthemum nummularium

This is a low-growing perennial up to 35cm high, which often grows in masses. The leaves are narrowly oval, 1–2cm long and white-hairy below, with two small, leaf-like stipules at each node. The bright golden-yellow flowers are 2–2.5cm across, with five petals and five uneven sepals, borne in loose clusters of up to eight flowers.

FLOWERING TIME June–September.

DISTRIBUTION Common and widespread over most of lowland Britain, but absent as a native plant from Ireland. It is most abundant in open calcareous habitats such as chalk downland, limestone grassland and calcium-rich dunes.

SIMILAR SPECIES **Hoary Rock-rose *H. oelandicum*** has smaller flowers up to 1.5cm across, very hairy leaves and no stipules. It is a rare plant confined to open limestone habitats in western England, Wales and western Ireland.

Common Rock-rose *Helianthemum nummularium*

WILD PANSY
Viola tricolor

Wild Pansy is a small annual or perennial up to 30cm tall. Pansies are very similar to violets (*see* p. 46) and closely related to them, but have divided leaf-like stipules, and the two side petals are held upwards so that the flower forms a 'face'. The leaves are oval and the stipules, pinnately lobed, are as large as the leaves. The flowers are yellow, blue-violet or both, and 1.5–2.5cm long with a short spur up to 5mm long.

FLOWERING TIME April–September.

DISTRIBUTION Frequent as an annual weed of gardens, fields and waste places throughout the region, though rare in western Ireland; also occurs as a perennial dune plant, ssp. *curtisii*, on western coasts.

SIMILAR SPECIES **Field Pansy *V. arvensis*** is similar, with smaller yellow flowers and petals shorter than the sepals.

Wild Pansy *Viola tricolor*

COMMON DOG-VIOLET
Viola riviniana

This small, hairless or slightly downy perennial is up to about 20cm tall when in flower. The heart-shaped leaves can be found both on the leafy flowering stems and on a central rosette of leaves. At the base of each leaf is a stipule with a fringe of hair-like lobes. The flowers are violet to purple, 2–2.5cm across and unscented, with the largest petal prolonged into a broad, blunt, furrowed spur that is paler than the petals and sometimes almost white.

FLOWERING TIME April–June.

DISTRIBUTION Common throughout Britain and Ireland in a wide variety of habitats, especially woodland, scrub and hedgerows, but also often in open grassland.

SIMILAR SPECIES There are many. One common, very similar species is **Early Dog-violet *Viola reichenbachiana***, which differs in having narrower leaves and stipules, and a straight, dark-coloured spur without a furrow at the end. It is common and widespread in Ireland and southern Britain, but largely absent from Scotland. **Sweet Violet *Viola odorata*** is downy-hairy, with fragrant flowers and rounded, shinier leaves that eventually become quite large. It is native or naturalized in most areas.

Common Dog-violet *Viola riviniana*

GARLIC MUSTARD
Alliaria petiolata

An erect, normally unbranched herbaceous perennial up to about 1m tall, Garlic Mustard has kidney- to heart-shaped, stalked leaves in a loose basal rosette and up the stem. The leaves smell strongly of garlic when crushed (they can be used in cooking, though they are a little strong for some tastes). The flowers are white, 3–5mm across, with four petals arranged in a cross (in common with all members of the mustard family); they are carried in an inflorescence that continues to elongate as the thin, erect fruits develop.

FLOWERING TIME April–June.

DISTRIBUTION A common and widespread plant almost everywhere except northern Scotland and western Ireland, in woodland, hedge-banks, gardens, churchyards and other semi-shady habitats.

Garlic Mustard *Alliaria petiolata*

CHARLOCK
Sinapis arvensis

Charlock is an erect, bristly-hairy annual up to 1m tall, with stalked, lyre-shaped lower leaves and unstalked, barely lobed upper leaves. The yellow flowers are 1.5–2cm across, four-petalled and borne in a dense terminal inflorescence. The fruits are 3–4cm long when mature, with the section containing the seeds about twice as long as the unseeded beak.

FLOWERING TIME May–August.

DISTRIBUTION Common and widespread throughout the region in open, disturbed habitats such as arable fields, roadsides and waste ground.

SIMILAR SPECIES White Mustard *Sinapis alba* is very similar to Charlock, but all its leaves are pinnately lobed, and the beak of the fruit is flat, sword-like and at least as long as the lower, seeded section. It is also widespread and common throughout most of the lowlands, though it is rare in both Ireland and Scotland.

Charlock *Sinapis arvensis*

Sea Stock
Matthiola sinuata

Sea Stock is an erect, hairy, biennial plant that is usually covered with yellowish or black glands. There is a distinctive rosette of greyish, wavy-edged leaves, from which an erect leafy flower stalk, up to 70cm high, is produced. The flowers are pinkish-purple, fragrant, four-petalled and each about 2–2.5cm across, held in a loose raceme. The fruits are long, narrow, up to 15cm long and sticky with yellowish or blackish glands.

FLOWERING TIME May–August.

DISTRIBUTION
Occurs rather uncommonly in coastal habitats such as sand dunes and clifftops, mainly in western Britain.

SIMILAR SPECIES
Hoary Stock
Matthiola incana
has unlobed entire leaves that are not covered with glands. It is a rare plant of cliffs and other coastal habitats in the south of England and Wales, but it is also widely cultivated in gardens and naturalized.

Sea Stock *Matthiola sinuata*.

WILD CABBAGE
Brassica oleracea

Wild Cabbage, the ancestor of cultivated cabbage, kale, cauliflower and other related vegetables, is one of the most distinctive of the yellow-flowered crucifers. It is a robust, hairless, greyish perennial, up to 80cm high, with woody-based stems bearing numerous leaf scars. The lower leaves are waxy, undulate and deeply lobed. The lemon-yellow flowers, each 2–3cm across, are borne in long, dense, conspicuous racemes, followed by long, narrow, cylindrical pods.

FLOWERING TIME April–June.

DISTRIBUTION Common around the coasts of Britain, especially in southern England and Wales, though it is not always clear where it is growing as a native plant.

Wild Cabbage *Brassica oleracea*

HAIRY BITTER-CRESS
Cardamine hirsuta

This small annual could be easily overlooked were it not for the fact that it is often an abundant garden weed. It has a basal rosette of pinnate leaves and erect, flowering stems up to about 30cm tall at the most. The flowers are white, four-petalled, very small (about 5–6mm across) and have four stamens; they are borne in loose, branched inflorescences.

FLOWERING TIME March–September.

DISTRIBUTION
Widespread and very common everywhere in dry, open habitats including gardens, arable fields and waste places.

SIMILAR SPECIES
Wavy Bitter-cress *Cardamine flexuosa* is similar, though generally taller, with wavy stems and flowers with six stamens. It is common throughout the region in damper disturbed habitats.

Hairy Bitter-cress *Cardamine hirsuta*

COMMON SCURVY-GRASS
Cochlearia officinalis

This sprawling or erect, hairless biennial or perennial is up to 40cm tall. The leaves are fleshy; the basal leaves are long-stalked, rounded and heart-shaped, and the stem leaves clasp the stem. The flowers are four-petalled, 8–10mm across, and usually white, with the petals 2–3 times as long as the sepals.

FLOWERING TIME May–August.

DISTRIBUTION Widespread and common around all coasts (except south-east England) and in mountain habitats, perhaps as a separate species.

SIMILAR SPECIES Danish Scurvy-grass *Cochlearia danica* is smaller, up to 20cm tall, with lobed, ivy-shaped, stalked stem leaves, and smaller (4–5mm), pale purple flowers. It is common in similar coastal habitats throughout the region, and now also along salted roadsides inland.

Common Scurvy-grass *Cochlearia officinalis*

CUCKOOFLOWER
Cardamine pratensis

Cuckooflower is an erect, hairless perennial up to 60cm tall, with a basal rosette of pinnate leaves that have rounded, coarsely toothed leaflets, and pinnate stem leaves with narrow, untoothed leaflets. The flowers vary in colour from lilac-pink to almost white, are up to 2cm across, and have four slightly notched petals and yellow anthers. The fruits are narrowly cylindrical and are about 4cm long when mature. Also known as Lady's-smock, this is a key food-plant of the Orange-tip Butterfly.

FLOWERING TIME April–June.

DISTRIBUTION Common and widespread throughout Britain and Ireland in open, damp habitats, particularly wet meadows and marshes, but also roadsides and open, damp woodland.

SIMILAR SPECIES The much rarer **Coralroot,** or **Coralroot Bittercress,** *Cardamine bulbifera* differs in having deeper-pink flowers, trifoliate upper leaves and shiny red bulbils in the leaf axils. It is a rare plant of southern England woodland. **Water-cress** *Nasturtium officinale* could be confused with Cuckooflower, but has smaller, clustered white flowers and the peppery-tasting leaves have a much larger end leaflet. It is common everywhere in watery habitats such as streams and ponds.

Cuckooflower *Cardamine pratensis*

BELL HEATHER
Erica cinerea

This dwarf evergreen, much-branched subshrub is up to 70cm tall. The leaves are small, dark green and up to 7mm long. They are very narrow with inrolled margins that cover the underside, and borne in groups of three on the stem and in leafy side-shoots. The flowers are bell- to urn-shaped, bright pinkish-purple, 4–6mm long and borne in loose whorls collectively making up a terminal inflorescence.

FLOWERING TIME July–September.

DISTRIBUTION Widespread and generally common over most of Britain and Ireland, with the exception of a few lowland areas, on acid moors, heaths and mountains, and in open woodland. In many situations it is abundant to dominant.

SIMILAR SPECIES **Cross-leaved Heath *E. tetralix*** is similar in form, but has greyish, hairy leaves in whorls of four. The pale pinkish flowers are hairy and grouped in a tight terminal, drooping, umbel-like cluster. It is common and widespread in damp, acidic places.

Bell Heather *Erica cinerea*

Bell Heather *Erica cinerea*

Heather, or **Ling**, *Calluna vulgaris* is similar in form, but has tiny, appressed scale-leaves, and very small, open, cup-shaped, four-petalled, pink flowers in tight, terminal, narrow inflorescences. It is common or abundant in many dry and damp acid habitats throughout the region, frequently dominating over large areas.

BILBERRY
Vaccinium myrtillus

This low-growing, deciduous shrublet is up to about 60cm high. It has distinctly four-angled green twigs and many branches. The finely toothed, oval leaves are alternate and about 2cm long. The greenish-pink or red flowers are urn-shaped and pendulous, about 6mm long, and scattered in ones and twos towards the ends of the twigs. The fruits are the familiar edible bilberry or whortleberry – black with a bluish bloom and spherical with a flattened end.

FLOWERING TIME April–June, with the fruit ripening from late July–September.

DISTRIBUTION Widespread and generally common on heaths and moors or in open woodland and acid soils in Britain, especially in western and upland areas.

SIMILAR SPECIES Bog Bilberry *V. uliginosum* is similar, but has round, brown twigs, greyish, untoothed leaves, and smaller, pale pink flowers. It occurs in damp upland areas, mainly in Scotland, and is absent from Ireland.

Bilberry *Vaccinium myrtillus*

ROUND-LEAVED WINTERGREEN
Pyrola rotundifolia

This low-growing herb has a rosette of glossy, round, long-stalked leaves, with the blade shorter than the petiole. An erect raceme of pendulous white flowers up to 30cm high is produced from the centre of the rosette. The wide-open flowers are flat (not bell-shaped), about 12mm across, with five petals; the solitary, S-shaped style is long (about 1cm) and protrudes out beyond the petals.

FLOWERING TIME July–September.

DISTRIBUTION
Scattered among marshes, dune-slacks, mountain ledges and other damp, usually calcareous habitats throughout Britain and Ireland, although absent from many areas.

SIMILAR SPECIES
Common Wintergreen *Pyrola minor* is smaller, with shorter-stalked leaves and bell-shaped flowers without the protruding style. It occurs throughout Britain and Ireland in rocky places and woods with acid soil.

Round-leaved Wintergreen *Pyrola rotundifolia*

WELD
Reseda luteola

This is a tall, hairless, erect biennial up to 130cm tall. Both the rosette leaves and the stem leaves are simple, narrow, unlobed, wavy-edged and up to 8cm long. The stem ends in a narrow, cylindrical inflorescence of small, greenish-white flowers, each with four petals and roughly 5mm across. Under a lens, it can be seen that three petals are lobed, while one is entire. It is also known as Dyer's Rocket.

FLOWERING TIME June–August.

DISTRIBUTION Common throughout the lowlands of Britain and Ireland, especially on calcareous soils, in grassland, waste ground, cultivated areas and other open habitats.

SIMILAR SPECIES **Wild Mignonette *R. lutea*** is similar, but its leaves are pinnately lobed, and the inflorescence is usually broader, paler and less dense. It is common in similar habitats, but more confined to calcareous places, and rare in western Britain and Ireland.

Weld *Reseda luteola*

PRIMROSE
Primula vulgaris

One of the most familiar and best loved of all spring wild flowers, Primrose is a hairy perennial with a rosette of oval leaves, the blades of which taper gradually towards the bases. The flowers arise singly on unbranched, leafless, hairy stems up to about 10cm tall. The flowers themselves are pale yellow, often with orange markings in the centre, about 3cm across, and fragrant. Primrose was formerly used as a medicinal plant (rarely so nowadays). It is widely grown in gardens, especially in different colour forms and as a hybrid, which may sometimes naturalize, for example in churchyards.

FLOWERING TIME March–May.

DISTRIBUTION Common and widespread across Ireland and the whole of Britain, especially in the south west, in woods, hedge-banks, churchyards, roadsides, scrub, meadows and grassland, usually on relatively heavy soils or in humid situations.

Primrose *Primula vulgaris*

COWSLIP
Primula veris

Cowslip is a familiar and attractive herbaceous spring flower. The leaves are oblong to oval, all in a basal rosette, and differ from those of Primrose in that they taper sharply rather then gradually to a winged stalk. A robust, leafy stem arises from the centre of the rosette, bearing an uneven umbel of 10–30 flowers not strongly turned to one side. The flowers are partly tubular, orange-yellow with deeper markings in the throat and about 8–10mm across.

FLOWERING TIME April–May.

DISTRIBUTION Widespread in southern Britain and Ireland, rarer further north, in open, calcareous habitats such as downland, meadows, dunes and open woodland.

SIMILAR SPECIES Oxlip *Primula elatior* is similar, but has paler green, less wrinkled leaves and pale, clear yellow, primrose-coloured flowers in an umbel that is strongly held to one side. It is restricted to woodland in East Anglia. The commonly occurring natural hybrid

Meadow of Cowslip *Primula veris*

Cowslip *Primula veris*

between Primrose and Cowslip **P. x polyanthema** looks like an
Oxlip, but the umbel is not one-sided, and the yellow flowers have
streaks of orange in the centres. This hybrid may occur anywhere
within the range of the parents.

Yellow Loosestrife
Lysimachia vulgaris

This creeping herbaceous perennial gives rise to erect, leafy, flowering stems up to 1.5m high, though usually less. The pointed, oval leaves are about 4–10cm long and borne in opposite pairs or in whorls of three to four. The bright yellow flowers are 1.5–2cm across and are borne in broad, leafy, pyramidal, terminal panicles. The edges of the sepals are orange and hairy; the margins of the petals are hairless.

FLOWERING TIME April–June.

DISTRIBUTION
Common in fens, marshes and damp, open woodland throughout Britain and Ireland, except the far north.

SIMILAR SPECIES
Dotted Loosestrife *Lysimachia punctata* is similar but evergreen, with a central red eye to the flowers and hairy-edged petals. It was introduced from the USA and widely naturalized.

Yellow Loosestrife *Lysimachia vulgaris*

SCARLET PIMPERNEL
Anagallis arvensis

Scarlet Pimpernel is a low-growing, prostrate, sprawling, annual plant with square stems up to 20cm high. The oval, unstalked leaves, usually in opposite pairs, are about 1cm long. The solitary flowers are flat to saucer-shaped on long stalks from the leaf axils, 10–15mm across, with five orange-scarlet petals, densely fringed with minute hairs. Plants with blue flowers may be found and these are either blue versions of Scarlet Pimpernel, or Blue Pimpernel *A. arvensis* ssp. *caerulea*, though the differences are minor.

FLOWERING TIME June–September.

DISTRIBUTION Common and widespread throughout most of Britain and Ireland in open, dry habitats such as arable land, dunes and waste ground.

Scarlet Pimpernel *Anagallis arvensis*

BITING STONECROP
Sedum acre

Biting Stonecrop is a low-growing, hairless succulent that produces erect, leafy shoots, with or without flowers. The leaves are fleshy, egg-shaped, appressed to the stem and peppery to the taste. The yellow, five-petalled flowers are about 1cm across and are borne in open-branched, few-flowered inflorescences on stalks up to 12cm tall.

FLOWERING TIME June–July.

DISTRIBUTION Widespread and common in open habitats on thin soils in places such as sand dunes, shingle, heaths, walls and even the roofs of buildings throughout Britain and Ireland, though least common in highland regions.

SIMILAR SPECIES White Stonecrop *Sedum album* is similar to Biting Stonecrop, but more sprawling and with white or pink-tinged flowers. It grows in similar habitats.

Biting Stonecrop *Sedum acre*

Opposite-leaved Golden-saxifrage
Chrysosplenium oppositifolium

A low-growing, sparsely hairy, mat-forming herb with spreading, leafy, non-flowering shoots, this species has erect flowering shoots bearing pairs of opposite, rounded leaves with wedge-shaped bases. The flowers are small, 3–4mm across, greenish-yellow and without petals, and are clustered together in terminal groups with broad yellowish bracts.

FLOWERING TIME April–June.

DISTRIBUTION Common in damp, shady places, especially by springs and streams, and on damp upland rocks, throughout the British Isles, except for the driest parts.

SIMILAR SPECIES Alternate-leaved Golden-saxifrage *Chrysosplenium alternifolium* is similar, but has alternate stem leaves rather than opposite pairs of leaves. The most obvious distinctions are the long-stalked, kidney-shaped basal leaves and the lack of creeping, leafy, non-flowering stems. It occurs in similar habitats, but favours more base-rich situations and is generally less common. It is absent as a native plant from the whole of Ireland.

Opposite-leaved Golden-saxifrage *Chrysosplenium oppositifolium*

MEADOWSWEET
Filipendula ulmaria

A familiar sight in summer, when its frothy masses fill fields and riversides, this is a tall perennial up to 1.2m high. The lower leaves are stalked, up to 50cm long, and pinnately divided into smaller leaflets roughly the shape of elm leaves, with much smaller leaflets between them. The upper leaves are similar but smaller. There are large, rounded stipules at the base of each leaf. The small, creamy flowers are held in large, untidy, umbel-like panicles, forming a conspicuous dense, fragrant mass. In past times, Meadowsweet was strewn on the floors of barns, churches and even houses to sweeten the air, and it was also used medicinally as a pain reliever.

FLOWERING TIME June–September.

Meadowsweet *Filipendula ulmaria*

DISTRIBUTION Common and very widespread throughout Britain and Ireland in damp, open habitats such as wet meadows, marshes and riversides, often forming dense, dominant masses.

SIMILAR SPECIES A rather similar species, easily confused with Meadowsweet, is **Dropwort *F. vulgaris***. It is less tall, with pinnate leaves more evenly divided into many small, toothed leaflets; the inflorescence consists of fewer larger flowers (1–2cm across, compared with less than 1cm in Meadowsweet), often tinged with red or pink. It has a more restricted distribution, mainly in lowland areas on calcareous soils, and is very rare in Ireland.

WILD STRAWBERRY
Fragaria vesca

Wild Strawberry is a small perennial, up to 30cm tall, characterized by its long, arching, rooting runners that allow it to form large patches and colonize new areas. The leaves are bright glossy green above, paler below, hairy and divided into three unstalked leaflets. The flowers are white, five-petalled and up to 2cm across; they grow in loose, few-flowered, branched clusters. The fruits, when ripe, are bright red and fleshy, like miniature garden strawberries and just as edible. In fact, they are not true fruits but the swollen stem tip, which has the tiny fruits in pits all over its surface.

FLOWERING TIME April–July.

DISTRIBUTION Widespread and common throughout the area in woodland, scrub, roadsides and meadows, normally on neutral or calcareous soils.

SIMILAR SPECIES Hautbois Strawberry *F. moschata* is larger and taller, with duller green, stalked leaflets. It is much rarer in Britain and Ireland. The rather similar **Barren Cinquefoil *Potentilla sterilis*** has bluish-green leaflets and similar flowers, but with noticeable gaps between the petals, so that the sepals are visible, and no fleshy red fruits. It is common throughout the region in woodland and hedgerows, flowering in March–May.

Wild Strawberry *Fragaria vesca*

SILVERWEED
Potentilla anserina

Silverweed is a low-growing, creeping perennial, with long, rooting, red runners, bearing its leaves in rosettes. The leaves are up to 12cm long and pinnately divided into 7–12 pairs of toothed, oblong leaflets; these are rather silvery above and very silky-silvery below, a feature that can be readily seen from a distance. The flowers are long-stalked, solitary and 1.5–2cm across, with five yellow petals, rather like buttercups (though one way the *Potentilla* species can be distinguished from buttercups is by the extra ring of calyx – the epicalyx – visible outside the normal ring of sepals).

FLOWERING TIME May–August.

DISTRIBUTION Widespread and common throughout Britain and Ireland in open, often trampled habitats such as tracks, car parks, gateways and waste ground. The scientific name *anserina* indicates a relationship with geese, and the plant was indeed once common wherever geese trampled and grazed.

Silverweed *Potentilla anserina*

TORMENTIL
Potentilla erecta

This much-branched, low-growing, perennial herb is up to about 30cm high. The leaves are trifoliate, but with two leafy stipules at the base so that they may appear digitate; the lower leaves are stalked, while the upper ones are stalkless. The bright yellow flowers are about 1cm across, with only four petals (almost all other cinquefoils, *Potentilla* species, have five petals) loosely grouped into clusters on long stalks.

FLOWERING TIME May–September.

DISTRIBUTION Widespread and common on more acid soils in a variety of open habitats such as heathland, grassland, dunes and roadsides.

SIMILAR SPECIES Creeping Cinquefoil *Potentilla reptans* has larger, five-petalled flowers, with five to seven leaflets per leaf and long red runners that root at the nodes. It is a common and invasive weed throughout Europe except in the far north.

Tormentil *Potentilla erecta*

WOOD AVENS
Geum urbanum

This erect, downy, branched perennial is up to 60cm high. The basal leaves are irregularly pinnately lobed, with a large, lobed end leaflet, and two smaller perpendicular, leafy stipules at the base. The stem leaves are unstalked, trifoliate or undivided. The bright yellow flowers are five-petalled, about 1–1.5cm across, with five triangular sepals directly underneath and in between the petals, and are borne on long stalks in loose-branched clusters. The reddish fruit is a spherical head of many hairy achenes, each equipped with a hook for attaching to any likely dispersal agent (especially dogs!). It is also known as Herb Bennet. It is often considered a weed, especially because its strong, deep roots and large number of seeds make it a rather difficult plant to eradicate from a garden once it manages to get a foothold there.

FLOWERING TIME May–September.

DISTRIBUTION Common in open or semi-shaded situations, such as light woodland, waste ground, roadsides and gardens throughout both Britain and Ireland.

Wood Avens *Geum urbanum*

SAINFOIN
Onobrychis viciifolia

Sainfoin is an erect, downy perennial up to 60cm tall. The leaves
are pinnate, with 6–12 pairs of narrow, pointed, oblong leaflets,
mainly up the stem, with papery stipules at the base of each. The
flowers are pink with deeper purple veins, grouped into roughly
conical, dense terminal inflorescences.

FLOWERING TIME June–August.

DISTRIBUTION
Locally native, but
widely introduced
as a fodder crop and
often naturalized.
It is confined mainly
to lowland Britain,
on calcareous
open habitats
such as downland,
roadsides and
areas of former
cultivation. It is
absent from Ireland.

SIMILAR SPECIES
Crown Vetch
Securigera varia
has pink flowers and
pinnate leaves, but
the flower heads
are more globular,
and the leaflets are
broader and shorter.
It is naturalized or
planted in rough,
grassy places
throughout all of
lowland Britain.

Sainfoin *Onobrychis viciifolia*

73

Common Bird's-foot-trefoil
Lotus corniculatus

This is a sprawling or erect, low-growing, more or less hairless perennial up to 40cm tall, with solid stems. The leaves are mainly on the stem, and have five oval to rounded leaflets – the lowest two resembling stipules, though actually there are two tiny stipules, too. The flowers are borne in terminal clusters of two to seven. They are yellow or orange, up to 1.5cm long and often streaked with red (giving rise to an alternative English name of Bacon and Eggs).

FLOWERING TIME June–September.

DISTRIBUTION Widespread and common throughout Britain and Ireland in open, grassy habitats – meadows, pastures, downland, dunes, roadsides – except on very acid soils.

SIMILAR SPECIES A widely sown variety, **L. corniculatus var. sativus**, is taller, with solid stems and plain yellow flowers. It is a common component of roadside seed mixes, though not native to Britain. **Greater Bird's-foot-trefoil L. pedunculatus** is more robust, up to 90cm tall and usually hairy, with hollow stems; a small though distinctive difference is that in this species the calyx teeth are strongly reflexed in bud, whereas in Common Bird's-foot-trefoil they are erect. It is widespread in most of the region in damper grassy places.

Common Bird's-Foot-Trefoil *Lotus corniculatus*

HORSESHOE VETCH
Hippocrepis comosa

Horseshoe Vetch is a creeping, hairless, perennial, herbaceous plant up to 40cm high. The leaves are pinnate, with up to 10 pairs of elliptic-to-oblong leaflets plus a terminal leaflet, and two narrow pointed stipules at the leaf base. The yellow pea-flowers are held in terminal clusters of up to 10 flowers on long stalks. The fruits are long and sinuous, constricted into horseshoe-shaped segments and held in a star shape. It is an important butterfly food plant, favoured by a wide range of species.

FLOWERING TIME May–July.

DISTRIBUTION Common only in England, especially in the south, and virtually absent further north or west in the British Isles. It can be found most frequently in dry, calcareous grassland, such as chalk downland and cliff-tops, and also appears as groundcover in gardens where the soil is chalky.

Horseshoe Vetch *Hippocrepis comosa*

COMMON RESTHARROW
Ononis repens

Common Restharrow is a spreading, rhizomatous, glandular-hairy perennial up to 60cm high. It is woody at the base, with the stems hairy all round. The toothed leaves, either simple or trifoliate, are stickily hairy and strong-smelling, with clasping stipules. The pink, pea-like flowers, in loose leafy racemes, are up to 2cm long.

FLOWERING TIME June–September.

DISTRIBUTION Widespread and locally common in well-drained, open sites such as downland, dunes, cliff-tops and field margins throughout Europe, except for northern Scotland and Northern Ireland.

SIMILAR SPECIES **Spiny Restharrow *O. spinosa*** is similar, but it is normally spiny (Common Restharrow may occasionally be spiny), the stems are hairy on two sides only, the leaflets are narrower and more pointed, and the flowers are deeper pink. It is widespread in similar habitats, especially on heavier soils.

Common Restharrow *Ononis repens*

TUFTED VETCH
Vicia cracca

A weakly scrambling perennial up to 1m or so tall, and occasionally up to twice that height, Tufted Vetch climbs with the aid of branched tendrils that attach themselves to other plants. The leaves are pinnate with up to 14 pairs of narrowly oblong, downy leaflets, each ending in a branched tendril. The flowers are held in dense, long-stalked, narrow, one-sided racemes up to 10cm long; each raceme bears 10–40 attractive bluish-purple flowers.

FLOWERING TIME June–August.

Tufted Vetch *Vicia cracca*

DISTRIBUTION
Common and widespread in most of the region in many rough, grassy habitats such as meadows, road verges, dunes and scrub.

SIMILAR SPECIES
Fine-leaved Vetch
Vicia tenuifolia is similar, but has larger flowers (1.2–1.8cm long) and narrower leaflets. It has been introduced and occasionally naturalized in Britain, and is widespread throughout much of mainland Europe.

MEADOW VETCHLING
Lathyrus pratensis

A weakly scrambling, downy perennial, Meadow Vetchling has angled, slightly winged stems reaching up to 1m in height. The leaves each consist of two opposite, greyish-green, narrow, pointed leaflets up to 3cm long, with a long, branched tendril between them; at the base of each leaf there are two leafy, arrow-shaped stipules up to 2cm long. The flowers are 1–1.6cm long and grouped into long-stalked, terminal inflorescences of up to 12 flowers. In the autumn, the yellow flowers turn into shiny, black seed pods that resemble peapods. A herbal tea for use as a cough remedy was once made from the leaves.

FLOWERING TIME May–August.

DISTRIBUTION
Common and very widespread throughout the region growing through hedges or in rough, grassy situations, such as roadsides, waste ground, woodland, pastures and meadows, particularly where the vegetation is tall or the ground is a little damp.

Meadow Vetchling *Lathyrus pratensis*

White Clover
Trifolium repens

This is a low-growing, creeping perennial up to about 20cm tall, with rooting stems. The leaves are typical of most clovers – trifoliate (with three roughly equal leaflets) – each roughly oval, up to 1cm long and normally with a white, arrow-shaped mark in the centre. They are held erect on the ends of long stalks. The flowers are white (turning brownish as they age) and small, but produced in dense terminal, fragrant, globular clusters that are notably attractive to bees. The species is frequently planted as a fodder crop, especially for its ability to raise soil nitrogen levels.

FLOWERING TIME June–September.

DISTRIBUTION Very common and widespread in many short, grassy habitats, such as pastures, meadows, roadsides and lawns, on virtually all soil types except very acid ones.

SIMILAR SPECIES Red Clover *T. pratense* is broadly similar in form, but has red to pink flowers in oval heads, borne on taller leafy stems, including leaves just below the flower head. It has roughly similar habitats and distribution. **Strawberry Clover *T. fragiferum*** is very similar to White Clover, but has pinkish, globular flower heads that become swollen, downy, spherical, pinkish heads in fruit (vaguely resembling strawberries, hence the common name). It is locally common in lowland Britain, especially near the sea, and usually on damp, heavy soils.

White Clover *Trifolium repens*

PURPLE-LOOSESTRIFE
Lythrum salicaria

This downy perennial has a creeping rootstock from which many erect stems – up to about 1m tall – arise. The leaves are oval, long-pointed, stalkless and up to 4cm long; they are usually borne in pairs, or occasionally in threes lower down the stem. The flowers are magenta to reddish-purple, six-petalled and borne in dense whorls towards the top of the stem, collectively forming a conspicuous long inflorescence. The sticky seeds adhere to the feet of birds, mammals and fishermen, and are spread in this way.

FLOWERING TIME June–September.

DISTRIBUTION
Widespread and moderately common in damp, marshy places, especially alongside streams, rivers and lakes, throughout the UK, although it is less common in Scotland. It can form dense, colourful stands over large areas that provide a strong attraction for bees, moths and butterflies.

Purple-loosestrife *Lythrum salicaria*

Rosebay Willowherb
Chamerion angustifolium

This tall, erect, robust, patch-forming perennial reaches up to 1.4m tall in favourable conditions. The leaves are narrow and arranged alternately (actually in a spiral, though this can be hard to see) all the way up the stem. The large (2–3cm across) rose-purple flowers are gathered into a long, conspicuous, tapering inflorescence, on which they open from the base upwards. The long central stigma is four-lobed and the stamens bend downwards. It is a strikingly attractive plant, often occurring in huge drifts of colour.

FLOWERING TIME July–September.

DISTRIBUTION Common in open habitats such as waste ground and forest clearings throughout the area, except western Ireland.

SIMILAR SPECIES Great Willowherb *Epilobium hirsutum* is rather similar but softly hairy all over, with opposite leaves lower down and leafy bracts among the flowers; it produces a less dense inflorescence. It is common in damp, grassy places and on the margins of lakes and rivers.

Rosebay Willowherb *Chamerion angustifolium*

Common Milkwort
Polygala vulgaris

This is a small, low-growing, hairless perennial up to 20cm tall. The leaves are small, roughly ovate and all arranged alternately, generally getting larger up the stem. The flowers are unusual and distinctive: they are irregular, with three tiny outer sepals, two large, petal-like inner sepals and three small petals joined together with eight stamens into a pale, fringed tube. The flowers can be blue, pink or white; they are 5–8mm long and borne in terminal racemes of up to 40 flowers.

FLOWERING TIME May–September.

DISTRIBUTION Widespread and common in sunny habitats such as downland, meadows and dunes, usually on neutral or calcareous soils.

SIMILAR SPECIES Thyme-leaved Milkwort *Polygala vulgaris serpyllifolia* is very similar, but has at least some of the lower leaves in opposite pairs, and more deeply coloured flowers. It is widespread in similar habitats, but usually on more acid soils, including heathland. **Chalk Milkwort *P. calcarea*** has a definite rosette of leaves near the base, with the lower leaves larger than the upper, and usually

Chalk Milkwort *Polygala calcarea*

bright blue flowers. Chalk Milkwort grows more locally in calcareous grassland habitats in lowland England.

Common Milkwort *Polygala vulgaris*

WOOD SORREL
Oxalis acetosella

This is a low-growing, creeping, perennial herb with long-stalked, delicate, light green, trifoliate leaves. The solitary, open cup-shaped flowers, 1–2cm across, are white with pale lilac veins and five petals. The long, thin flower stalks are about the same length as, or longer than, the leaf stalks.

FLOWERING TIME April–May.

DISTRIBUTION Widespread throughout in moist, shady situations, particularly woods. Several other sorrels *Oxalis* species occur, but all are naturalized from South Africa or North America. They normally have similar trifoliate leaves, but the flowers are either yellow (such as **Least Yellow Sorrel *Oxalis exilis***) or pinkish-red (such as **Pink Sorrel *Oxalis articulata***).

Wood Sorrel *Oxalis acetosella*

Meadow Cranesbill
Geranium pratense

Meadow Cranesbill is an erect, hairy, perennial plant that is stickily glandular-hairy above. The basal leaves are deeply divided almost to the base into five to seven lobes, each one further lobed; the upper leaves are similar, but smaller and unstalked. The large flowers, 3–4cm across, are purplish-blue to clear sky-blue, with unnotched petals, and are borne in pairs.

FLOWERING TIME June–September.

DISTRIBUTION Widespread but local in meadows, roadsides and other grassy areas, especially on lime-rich soils, in England, Wales and south Scotland, but are rare or absent elsewhere. It is grown in gardens and occasionally naturalized outside its natural range.

SIMILAR SPECIES
Wood Cranesbill
G. sylvaticum is similar, but it has flowers that are more reddish-purple, slightly smaller and less widely open, and its leaves have broader lobes. It is locally common in northern Britain in meadows, roadsides and open woods, particularly in hilly areas.

Meadow Cranesbill *Geranium pratense*

HERB ROBERT
Geranium robertianum

Herb Robert is a much-branched, hairy, strong-smelling annual (though occasionally overwintering) plant, up to 40cm tall. The leaves are palmately divided into five lobes (lower leaves), or three lobes (upper stem leaves). The flowers are bright pink with five equal petals, up to 1.8cm across, and have orange anthers. All parts of the plant are aromatic.

FLOWERING TIME April–September.

DISTRIBUTION Common and widespread in open and semi-shaded habitats including cliffs, dunes, gardens, roadsides and open woodland throughout the region, though rare in the far north.

SIMILAR SPECIES Little Robin *Geranium purpureum* is similar, but smaller in all parts, with flowers up to 1.2cm across that have yellow anthers. It is uncommon in southern England and Ireland, growing in coastal and calcareous habitats.

Herb Robert *Geranium robertianum*

COMMON STORK'S-BILL
Erodium cicutarium

This low-growing, spreading, hairy or stickily hairy annual is up to about 30cm tall. The leaves are all pinnately divided, with each leaflet pinnately lobed (a useful distinction between cranesbills *Geranium* and storksbills *Erodium* is that cranesbills have palmately divided leaves with all veins originating from a central point, whereas storksbills have pinnately divided leaves with the veins originating along the length of the main vein). The flowers are pink to purple, with five petals, and about 1cm across; they grow in loose umbels of up to seven flowers.

FLOWERING TIME May–September.

DISTRIBUTION Widespread and common throughout the region on light soils, but most frequent in coastal habitats.

Common Stork's-bill *Erodium cicutarium*

HIMALAYAN BALSAM
Impatiens glandulifera

A tall, hairless, robust annual, Himalayan Balsam grows up to 2m tall in favourable conditions and has fleshy, reddish, angled stems. The leaves are narrowly oval with red teeth, and are arranged in groups of three up the stem. The flowers of balsams are distinctive: five-petalled and strongly asymmetric, with a wide lower lip, hood and distinct spur. In this species they vary from pale to deep pink, are up to 4cm long with a short, curved spur and have a strong, musky smell. The alternative name of Policeman's Helmet refers to the shape of the flower – like an old-fashioned police constable's helmet.

FLOWERING TIME July–October.

DISTRIBUTION Although not native (an introduction from the Himalayas), it is now widespread and often invasive in damp habitats, especially along riversides. The ridged, oblong capsules explode when ripe, scattering the seeds onto the soil or into water to be carried to new habitats.

SIMILAR SPECIES Touch-me-not Balsam *I. noli-tangere* is less tall, with alternate leaves and bright yellow flowers. It is a rare native in parts of Wales and north-west England, but widely naturalized elsewhere in damp woods and riversides. Flowering time is similar.

Himalayan Balsam *Impatiens glandulifera*

Ivy
Hedera helix

Ivy is an evergreen, woody, perennial climber, reaching up to 30m tall where there is a support to climb on, or forming a carpet over woodland floors and other shady habitats. The older stems are covered with sticky, suckering rootlets that rapidly attach the plant to any support. Juvenile leaves, on non-flowering stems, are palmately divided into three to five untoothed lobes, and are glossy green above; mature leaves, on flowering shoots, are roughly oval and usually unlobed. The yellowish-green flowers are small, with five sepals, petals and stamens gathered together into roughly spherical umbels 2–3cm across. The fruits are dull black, almost spherical berries, much favoured by birds.

FLOWERING TIME September–November.

DISTRIBUTION Very common in woods, rocky places, gardens and other habitats throughout the region, except those at high altitudes.

Ivy *Hedera helix;* at right, with Red Admiral Butterfly

HOGWEED
Heracleum sphondylium

Hogweed is a robust, roughly hairy, erect, branched biennial up to 2m tall, with hollow, ridged stems. The leaves are large and pinnate, with broad, lobed and toothed leaflets; the upper leaves have large, inflated, clasping bases. The flowers are white and borne in large umbels up to 15cm across; the petals of the outer flowers in particular are very unequal in length.

FLOWERING TIME June–September.

DISTRIBUTION
Widespread and common throughout the area in often nutrient-rich habitats including meadows, roadsides, farmyards and gardens.

SIMILAR SPECIES
Giant Hogweed *H. mantegazzianum* is similar, but much larger, with red-spotted stems up to 5m tall and flower umbels up to 50cm across. It has been widely naturalized from the Caucasus, especially along rivers, and is strongly phototoxic, with contact causing skin blistering in sunlight.

Hogweed *Heracleum sphondylium*

COMMON CENTAURY
Centaurium erythraea

A hairless, erect annual, branched or unbranched, and up to 40cm tall, Common Centaury has a basal rosette of roughly oval leaves, and a few pairs of stem leaves often bearing small clusters of flowers. The flowers are pink, five-petalled, unstalked and about 1.2cm across, and are borne in branched, roughly flat-topped clusters.

FLOWERING TIME June–September.

DISTRIBUTION
Common in open, grassy habitats such as sand dunes, downland and roadsides throughout the area, except in the far north.

SIMILAR SPECIES
Lesser Centaury
C. pulchellum is similar, but smaller in all parts, with flowers on short stalks (2–4mm long) and without a basal rosette. Lesser Century is uncommon and often coastal in England and Wales, and is rare in Scotland and Ireland.

Common Centaury *Centaurium erythraea*

95

Marsh Gentian
Gentiana pneumonanthe

This is an erect or ascending, hairless perennial up to 40cm tall. The leaves are blunt, narrowly linear and up to 4cm long, growing in opposite pairs on the stem only (not in a basal rosette). The flowers are large, trumpet-shaped and bright blue with spotted green stripes on the outsides; they are up to 4cm long and borne in terminal clusters of up to ten flowers, though normally fewer.

FLOWERING TIME September–November.

DISTRIBUTION Uncommon, growing in scattered sites across England and Wales, particularly in southern locations such as England's New Forest, in damp, acidic habitats such as wet heaths and dune slacks. It is rare or absent in Scotland and Ireland, and generally declining through habitat loss.

Marsh Gentian *Gentiana pneumonanthe*

BITTERSWEET
Solanum dulcamara

Bittersweet is a scrambling, downy-hairy (occasionally hairless) perennial reaching about 2m, though usually less. The leaves are alternate, stalked and roughly oval, often with two lobes at the base. The flowers are distinctive and borne in a lax, branching inflorescence, with each flower 1–1.5cm across; there are five narrow purple petals, which are often strongly reflexed, and a conspicuous cone of bright yellow stamens. The fruits are egg-shaped berries up to 1cm long that are a shiny red when ripe.

FLOWERING TIME June–September.

DISTRIBUTION Common generally in damp places such as open woods, riversides, lake margins and waste ground throughout the region, though rare in northern Scotland and western Ireland.

SIMILAR SPECIES **Black Nightshade *Solanum nigrum*** is similar, only with black, spherical berries, white petals and an annual, more erect growth form. It is common in England and Wales, and rare or absent elsewhere.

Bittersweet *Solanum dulcamara*

VIPER'S BUGLOSS
Echium vulgare

An erect, bristly biennial (occasionally perennial) up to 75cm tall, Viper's Bugloss has single or multiple stems covered in red-based, swollen bristles. The leaves are narrowly elliptical to strap-shaped, appearing both in a rosette and alternately up the stem. The flowers are funnel-shaped with unequal petals, bright blue when open but pinkish in bud, 1.2–1.8cm across, and borne in curving clusters, each arising from a bract on the stem. There are five stamens, of which four are long and protrude well beyond the petals.

FLOWERING TIME June–September.

DISTRIBUTION Common in England and Wales in dry, open habitats such as downlands, sand dunes and cliff-tops, but rare in Scotland and Ireland.

Viper's Bugloss *Echium vulgare*

FIELD BINDWEED
Convolvulus arvensis

This scrambling or creeping perennial is well known to gardeners. It has more or less hairless stems coming from thick, fleshy rhizomes. The leaves are alternate, arrow-shaped to oblong, stalked and up to 5cm long. The solitary flowers are funnel-shaped, up to 3cm across and variable in colour, but usually white, pink, or pink with five white stripes. Each flower is short-lived and opens only in daylight.

FLOWERING TIME June–September.

DISTRIBUTION A widespread, common and often abundant plant in waste ground, cultivated areas and open coastal habitats throughout the region, though rare in northern Scotland. It is frequent and persistent in gardens.

SIMILAR SPECIES **Sea Bindweed** *Calystegia soldanella* has flowers that are similar in shape and colour, but much larger (up to 4cm across), growing from plants that have fleshy, kidney-shaped leaves. It occurs in sandy and shingle habitats around most of the coasts of Britain and Ireland. **Hedge Bindweed** *C. sepium* is a larger version of Field Bindweed, with white flowers up to 5cm across, and a pair of large bracteoles enclosing the sepals. It is common throughout the area, except the far north and higher upland areas, in hedges, gardens, wood margins, fens and other habitats, with similar flowering times.

Sea Bindweed *Calystegia soldanella*

Field Bindweed *Convolvulus arvensis* (above and below)

COMMON COMFREY
Symphytum officinale

This robust, erect, bristly perennial, up to 1.5m tall, has stems that are winged strongly all the way between each internode. The leaves are large and oval-lanceolate, with the bases of the stem leaves running on down the main stems. The reddish-purple, yellow, pink or cream flowers are tubular or narrowly bell-shaped, with short, triangular, reflexed petal lobes.

FLOWERING TIME May–July.

DISTRIBUTION Common and quite widespread in damp, fertile places such as riversides, marshes and roadsides, though not common in the north and west.

SIMILAR SPECIES **Russian Comfrey *S*. x *uplandicum*** is a hybrid between Common Comfrey and **Rough Comfrey *S. asperum***. In many areas it is now more common than either parent. It is similar to Common Comfrey, but more bristly, with bluer-purple flowers and wings that only extend partway along each internode.

Common Comfrey *Symphytum officinale*

WATER FORGET-ME-NOT
Myosotis scorpioides

The forget-me-nots are familiar plants with attractive blue flowers. This species is one of the most frequent and conspicuous, with erect stems up to 30cm tall rising from creeping runners that bear alternate oblong to lanceolate leaves. The flowers are five-petalled, bright blue and up to 1cm across; they are borne in curved inflorescences.

FLOWERING TIME May–September.

DISTRIBUTION Common and widespread almost everywhere in damp habitats, avoiding strongly acidic situations.

SIMILAR SPECIES There are many similar species, some of which are hard to identify. **Wood Forget-me-not *Myosotis sylvatica*** is similar, but lacks creeping runners, is much more hairy and grows in drier, usually semi-shaded situations. **Field Forget-me-not *Myosotis arvensis*** is annual, with flowers that are a paler, slightly greyish-blue and smaller (up to 5mm across). It is common in dry, open habitats such as arable fields and sand dunes.

Wood Forget-me-not *Myosotis sylvaticas*

Water Forget-me-not *Myosotis scorpioides*

BETONY
Stachys officinalis

Betony is an erect, softly hairy perennial up to 60cm tall, with a basal rosette of long-stalked, oval to heart-shaped leaves, and several pairs of leaves on the stem, all with deep, blunt teeth. The flowers are asymmetrical, bright reddish-purple (occasionally white or pink) and up to 16mm long; they are clustered together in dense whorls forming a short, oblong spike.

FLOWERING TIME June–September.

DISTRIBUTION Common in pastures that are at least lightly grazed, open woods, hedge-banks, meadows and roadsides, and usually on drier, slightly acidic soils. It can be found throughout England and Wales, but it is uncommon in both Scotland and Ireland.

Betony *Stachys officinalis*

Hedge Woundwort
Stachys sylvatica

This is an erect, vigorous, unpleasant-smelling herb up to 80cm high, with bristly square stems. The leaves are opposite, stalked and ovate, with a heart-shaped base. Its inflorescence is a loose, terminal spike made up of whorls of wine-red, two-lipped flowers, patterned with white dots.

FLOWERING TIME July–September.

DISTRIBUTION Common and widespread throughout Britain and Ireland in semi-shaded places such as woodlands, hedge-banks and occasionally an invasive weed in gardens.

SIMILAR SPECIES
Marsh Woundwort
Stachys palustris
is similar, but with narrower, short-stalked or unstalked leaves, no strong smell and paler pinkish-purple flowers. It has a similar distribution and flowering time to Hedge Woundwort, but is found in damper, sunnier habitats.

Hedge Woundwort *Stachys sylvatica*

WILD THYME
Thymus polytrichus

A familiar mat-forming, low-growing, spreading perennial, Wild Thyme has flowering shoots up to 15cm tall (though usually less) that smell of thyme when crushed. Its stems are bluntly four-angled, and hairy only on two opposite faces. The leaves are small, narrowly oval and borne in opposite pairs. The flowers are pinkish-purple, 6–8mm long and held in dense terminal clusters.

FLOWERING TIME May–August.

DISTRIBUTION Wild Thyme is widespread and common throughout the area in dry, grassy and rocky places such as heaths, dunes and downland.

SIMILAR SPECIES Large Thyme *Thymus pulegioides* is larger, has more ascending stems and is less mat-forming, with longer and more interrupted flower spikes. The stems are hairy on the angles only. Within Britain it is confined to dry, calcareous habitats in lowland England and eastern Wales.

Wild Thyme *Thymus polytrichus* ssp. *britannicus*

COMMON HEMP-NETTLE
Galeopsis tetrahit

Common Hemp-nettle is an erect, branched, bristly hairy, square-stemmed, annual plant up to about 80cm high. Its swollen nodes are covered with sticky, red-tipped, glandular hairs. The leaves, in opposite pairs, are stalked and oval in shape, narrowing gradually into the stalk. The inflorescence consists of loose spikes made up of whorls of flowers that are pale to deep pink with a yellowish centre and deeper pink markings on the lower lip, though these markings never extend to the edges of the lower petals.

FLOWERING TIME July–September.

DISTRIBUTION
Widespread throughout Britain and Ireland, except the far north, in arable land, roadsides, fens and other habitats.

SIMILAR SPECIES
Bifid Hemp-nettle
Galeopsis bifida
is very similar to Common Hemp-nettle, but it has more extensive dark markings reaching to the edges of the lower petals, which are turned back. It also has similar habitats and distribution.

Common Hemp-nettle *Galeopsis tetrahit*

WHITE DEAD-NETTLE
Lamium album

White Dead-nettle's English name refers to the fact that it can look like a nettle but does not sting. It is a hairy, patch-forming perennial with creeping stolons, from which arise square stems bearing opposite pairs of nettle-shaped leaves. The flowers are white and asymmetrical, with a curved, hairy upper lip forming a hood, and they are grouped together into several well-separated whorls.

FLOWERING TIME April–November.

DISTRIBUTION Common in England, Wales and southern Scotland in grassy habitats, usually on relatively deep, fertile soils, such as those found in gardens and wood margins. It is uncommon in northern Scotland and most of Ireland.

SIMILAR SPECIES Yellow Archangel *Lamiastrum galeobdolon* is similar, but has leafy above-ground runners, bright yellow flowers marked with red, and narrower leaves. It is common in woods in England and Wales, and rare or introduced elsewhere.

White Dead-nettle *Lamium album*

Self-heal
Prunella vulgaris

A slightly hairy, patch-forming, low-growing perennial, Self-heal has creeping runners and erect or ascending square stems up to 20cm tall. The leaves are oval, in opposite pairs, usually untoothed and sometimes lobed at the bases. The flowers are blue-violet, up to 1.5cm long and have a curved upper hood-forming petal; they are grouped together into a dense, short terminal inflorescence with a pair of leaves immediately below it. Self-heal is sometimes confused with Bugle but is easily distinguished from it by the short terminal inflorescence, the pair of leaves below it and the curved upper petal (absent in Bugle). As the name suggests, Self-heal was once used as a medicinal plant to treat wounds and other conditions.

FLOWERING TIME June–November.

DISTRIBUTION Very common and widespread throughout the area in grassy habitats, usually on neutral or calcareous soil.

Self-heal *Prunella vulgaris*

Self-heal *Prunella vulgaris*

SIMILAR SPECIES Ground-ivy *Glechoma hederacea* has rather
similar flowers but is borne in much looser, leafy whorls, and
the leaves are rounded, with regular teeth. It is a strong, slightly
unpleasant-smelling plant that was once used for flavouring beer,
and is common and widespread in most of the area, flowering in
March–May.

BUGLE
Ajuga reptans

Bugle is a perennial herb with long, leafy, rooting runners, and short, erect, square stems that are hairy on two opposite sides only. The leaves are oval, slightly toothed and stalked in the basal rosette, but become stalkless in pairs up the stem. The flowers are violet-blue, up to 2cm long and asymmetric, with the stamens protruding well beyond the tiny upper lip; they are grouped together into leafy whorls forming a single, unbranched terminal inflorescence.

FLOWERING TIME April–June.

DISTRIBUTION Widespread and common almost throughout the area in damp meadows, woodland and roadsides. It is well used by many insects, including bees and butterflies, as a spring source of nectar.

Bugle *Ajuga reptans*

WATER-MINT
Mentha aquatica

This downy, erect perennial smells strongly of mint when crushed or bruised. Its leaves are oval, arranged in opposite pairs, and toothed and stalked. The flowers are small, 4–6mm across and lilac-pink; they are clustered into a dense terminal head, usually with one or two separated whorls of flowers lower down the stem. Water-mint was once widely used for both culinary and medicinal purposes.

FLOWERING TIME July–September.

DISTRIBUTION Common and widespread in damp and marshy habitats almost everywhere.

SIMILAR SPECIES There are many other mints, including a number of hybrids and naturalized garden varieties. **Corn Mint *M. arvensis*** is similar, but differs in having flowers only in whorls, with no terminal cluster. It is normally less common, though widespread in drier or seasonally wet habitats.

Water-mint *Mentha aquatica*

LADY'S BEDSTRAW
Galium verum

Lady's Bedstraw is a sprawling perennial with erect or ascending, more or less hairless, flowering stems. The leaves are narrow, less than 2mm wide and dark green above; they grow in whorls of 8–12 up the stem. The flowers are small, 2–3mm across, and golden-yellow with four pointed petals; they are borne in loose, many-flowered, branched inflorescences. When dried, the plant is pleasantly scented – it was once used to stuff mattresses.

FLOWERING TIME July–September.

DISTRIBUTION Common and widespread in grassland, dunes and roadsides, often in great abundance, throughout the area.

SIMILAR SPECIES **Crosswort *Cruciata laevipes*** has similar yellow flowers, but the broader leaves are carried in distinctive whorls of four, with the flowers nestling among them. It is common in England and Wales, and rare elsewhere.

Lady's Bedstraw *Galium verum*

RIBWORT PLANTAIN
Plantago lanceolata

A familiar weedy, rosette-forming perennial, Ribwort Plantain is made up of a rosette of many narrow lanceolate leaves, each strongly ribbed with about five roughly parallel ridges from which hairy, leafless, ridged flower stems arise. The flowers are small, 3–4mm across, and have pale yellowish protruding stamens; they are clustered into narrow, oblong terminal spikes.

FLOWERING TIME May–September.

DISTRIBUTION Common in most open habitats, such as roadsides and meadows, throughout the region.

SIMILAR SPECIES
Greater Plantain
P. major has broader, oval leaves, unfurrowed flower stems and long inflorescences; its habitats and flowering time are similar to Ribwort Plantain. **Hoary Plantain** *P. media* has a flat rosette of broad, short-stalked leaves, and long spikes of white flowers with conspicuous pink or purple stamen filaments. It is found most frequently in calcareous grassland.

Ribwort Plantain *Plantago lanceolata*

FOXGLOVE
Digitalis purpurea

Foxglove is a tall, erect, downy biennial (occasionally perennial) up to 1.5m tall, but usually less. After the first year it produces a rosette of narrowly oval leaves, followed in the second year by the leafy flowering stem. The purple flowers are tubular to bell-shaped, slightly pendulous, marked inside with dark spots on white and up to 5cm long; they are borne in long, tapering spikes with the flowers opening at the bottom first. Completely white flowers sometimes occur, and these may be abundant in certain populations.

FLOWERING TIME June–August.

DISTRIBUTION Common and widespread throughout the area in woodland clearings, rides, hedge-banks and other habitats, on dry, acid soils.

Foxglove *Digitalis purpurea* (both left and below)

GERMANDER SPEEDWELL
Veronica chamaedrys

A creeping, hairy perennial with ascending flower spikes up to 25cm tall, Germander Speedwell often forms large, sprawling patches. The stems have two distinct opposite lines of long, white hairs, and opposite pairs of oval, short-stalked, toothed leaves. The flowers are bright blue with a white eye and about 1cm across; they are borne in long pairs of stalked inflorescences that arise from the axils of the uppermost pair of leaves.

FLOWERING TIME April–July.

DISTRIBUTION Common and widespread throughout Britain and Ireland in a variety of sunny habitats such as woodland clearings, scrub, roadsides, gardens and waste ground.

SIMILAR SPECIES **Wood Speedwell *Veronica montana*** is similar in form, with large leaves up to 3cm long, but the stems are hairy all around and the flowers are smaller (up to 7mm across), paler lilac-blue in colour and grow in open, few-flowered spikes. It is widespread and locally common in old woodlands throughout the area except the far north. **Heath Speedwell *V. officinalis*** has smaller leaves (1–2cm long) and dense spikes of pale blue-lilac flowers. It is common in dry, heathy and grassy places throughout the region, flowering in May–August.

Wood Speedwell *Veronica montana*

Germander Speedwell *V. chamaedrys*

GREAT MULLEIN
Verbascum thapsus

This tall, robust, erect, usually unbranched, white-woolly biennial grows up to 2m tall. The lower leaves are large and elliptical to oblong, with narrowly winged stalks; the stem leaves become progressively smaller and shorter-stalked, with the wings running down the main stem. The flowers are bright yellow, five-petalled, 2–3.5cm across and not quite symmetrical; they have five stamens, of which the upper three have white-woolly stalks. The flowers grow in a dense, long, leafy, usually unbranched spike.

FLOWERING TIME June–August.

DISTRIBUTION Common and widespread almost throughout the region, except north-west Scotland, in dry, often calcareous places such as banks, waste ground and roadsides.

Great Mullein *Verbascum thapsus*

SIMILAR SPECIES Dark Mullein *V. nigrum* is usually shorter, with leaves dark green on the upper surface, not white-woolly; the flowers are similar, up to 2cm across, but all five stamens have conspicuous long, violet-purple hairs on the filaments, giving the impression of a purple centre to the flower. **Moth Mullein *V. blattaria*** has flowers similar to Dark Mullein's, but they are larger, up to 3cm across, solitary and widely spaced on the stem. Though normally yellow, they may also frequently be white. The species is naturalized in scattered, open locations such as roadsides and quarries in lowland England only.

Ivy-leaved Toadflax
Cymbalaria muralis

Although not a native plant in Britain, this 17th-century introduction from the mountains of southern Europe is now so common and widespread that it can be considered an honorary native. It is a trailing, slender-stemmed, hairless perennial with alternate, long-stalked, deeply lobed, rounded to heart-shaped leaves (often resembling ivy in shape), usually 2–3cm across. The flowers are typical toadflax-shaped, purplish-lilac to almost white, about 1cm long, spurred and solitary; they are produced on long stalks from the leaf axils.

FLOWERING TIME May–September.

DISTRIBUTION Widespread and common throughout the region, except north-west Scotland, on walls and cliffs.

Ivy-leaved Toadflax *Cymbalaria muralis*

COMMON TOADFLAX
Linaria vulgaris

Common Toadflax is an erect, greyish-green, leafy perennial up to 70cm tall. All its leaves are narrowly linear, one veined and up to 7cm long; they are numerous, growing mainly in whorls lower down and becoming alternate above. The flowers are asymmetrical, bright yellow with an orange centre, up to 2.5cm long and have a long, straight, downwards-pointing spur; they are borne in dense, cylindrical terminal inflorescences.

FLOWERING TIME July–October.

DISTRIBUTION Widespread and common in grassy habitats such as meadows, cultivated ground and roadsides in England, Wales and southern Scotland, but it is found much less frequently in Ireland and northern Scotland.

Common Toadflax *Linaria vulgaris*

COMMON EYEBRIGHT
Euphrasia nemorosa

The eyebrights are a distinctive group of plants, easy enough to identify as eyebrights, but very hard to identify to species level. Common Eyebright is used here as an illustration of the type, but there are many similar species. All are partial parasites on grasses. This species is a short annual plant with many ascending branches, up to 20cm high, bearing pairs of small, oval, sharp-toothed leaves. The flowers are small, irregular, about 1cm long and white, sometimes with bluish markings; they are borne in loose, leafy inflorescences. It was once used in treating eye complaints, hence its name.

FLOWERING TIME July–September.

DISTRIBUTION Common and widespread in grassy habitats with moist and chalky soils in southern Britain and in parts of Ireland, but rare elsewhere. Other species of the group also occur throughout the region.

Common Eyebright *Euphrasia nemorosa*

Common Broomrape
Orobanche minor

The broomrapes are a group of wholly parasitic species with no green colour and simple, scale-like leaves. Many are specific parasites on one host species, or group of species, within a large range of hosts, often in the legume family and crop plants. It can be very helpful to identify the host when identifying a broomrape. Common Broomrape is generally the most widespread species. It consists of one (or more) erect, unbranched, yellowish or reddish stem terminating in a spike of flowers; these are asymmetrical, two-lipped and normally yellowish with purple veins, with the back of the upper lip smoothly curved. The large, two-lobed stigma is usually yellow, more rarely red.

FLOWERING TIME June–September.

DISTRIBUTION Common and widespread in Wales and England in meadows, cultivated land and roadsides; rare elsewhere.

SIMILAR SPECIES Ivy Broomrape *O. hederae* is similar, but its flowers are smaller, whiter, more spread out and have pointed corolla lobes; it is parasitic on Ivy, and mainly southern in Britain and Ireland.
Knapweed Broomrape *O. elatior* is up to 75cm tall with yellowish-brown stems, yellow flowers tinged purple, and yellow stigma lobes. It is parasitic on knapweeds *Centaurea* spp. in calcareous grassland in southern and eastern England.

Common Broomrape *Orobanche minor*

127

YELLOW-RATTLE
Rhinanthus minor

An erect, almost hairless annual, Yellow-rattle has black-spotted stems up to 50cm tall that bear pairs of narrowly oblong, toothed leaves. The flowers have a yellow, two-lipped corolla and a very flattened calyx that inflates to become bladder-like in fruit (the names Yellow-rattle and also Hay-rattle refer to the way in which the seeds rattle inside this inflated calyx). The flowers are clustered into a leafy terminal inflorescence.

FLOWERING TIME May–August.

DISTRIBUTION Widespread and moderately common throughout the area in grassy habitats, especially hay meadows and pastures. It is a partial parasite on grasses.

Yellow-rattle *Rhinanthus minor*

COMMON COW-WHEAT
Melampyrum pratense

Common Cow-wheat is an erect, usually hairless, branched annual plant up to about 50cm high. The leaves are narrowly oval to linear, more or less unstalked and in opposite pairs. The inflorescence is a loose, rather one-sided spike of pale yellow, tubular flowers, with the tube, up to 17mm long, almost closed at the end by the petal lobes.

FLOWERING TIME May–September.

DISTRIBUTION Widespread and generally common throughout, except in eastern England, in woods, heaths and grasslands, particularly on acid soils.

SIMILAR SPECIES
Small Cow-wheat
Melampyrum sylvaticum is generally smaller, with deeper yellow flowers that are shorter (8–10mm long) and more open at the mouth. It is rare and occurs only in north Britain and Northern Ireland in woodlands and other shady places. Both species are partial parasites on other plants.

Common Cow-wheat *Melampyrum pratense*

MARSH LOUSEWORT
Pedicularis palustris

This is an erect, almost hairless, single-stemmed but branched annual that becomes roughly pyramidal in shape as it matures. The leaves usually alternate (occasionally opposite) and are roughly oblong in outline but deeply pinnately divided into lobes. The flowers are pinkish-purple in loose spikes, with a distinctly two-lipped corolla about 2–2.5cm long. It is also known as Red Rattle.

FLOWERING TIME May–September.

DISTRIBUTION
Widespread and common almost throughout the area, though declining in damp habitats such as fens, bogs and wet meadows.

SIMILAR SPECIES
Lousewort *P. sylvatica* is similar, but perennial, shorter and multi-stemmed, with paler pink flowers that have a single tooth on each side of the upper lip (compared to two on each side in Marsh Lousewort). It has similar habitats and distribution.

Marsh Lousewort *Pedicularis palustris*

HAREBELL
Campanula rotundifolia

Harebell is a slender, erect or ascending, almost hairless perennial with creeping stolons, reaching to 40cm tall. The long-stalked basal leaves are round to heart-shaped (hence the scientific name *rotundifolia,* meaning round-leaved), while the stem leaves are quite different, narrowly linear and unstalked. The flowers are bell-shaped, 1.5–2cm long, pale blue and erect in bud, then drooping; they are borne in loose, open, few-flowered panicles. Harebell is known as Bluebell in Scotland.

FLOWERING TIME July–September.

DISTRIBUTION Common and widespread in dry, grassy habitats in most of the region, though curiously almost absent from south-west England and south-east Ireland.

Clustered Bellflower
Campanula glomerata

SIMILAR SPECIES Clustered Bellflower *C. glomerata* is a more robust, erect plant up to 30cm tall, with roughly hairy, ovate leaves and deep blue to violet flowers clustered in a tight, more or less terminal head. It grows in calcareous grassy areas, and is virtually confined to England within the area. **Creeping Bellflower *C. rapunculoides*** produces strong-growing, erect stems up to 70cm tall from creeping stolons, and often forms dense patches. The large, cup-shaped, drooping blue flowers, with reflexed sepals, are borne in a long, thin, one-sided inflorescence. It is commonly naturalized almost throughout Britain, and very rare in Ireland.

Harebell *Campanula rotundifolia*

MOSCHATEL
Adoxa moschatellina

Moschatel is a small and easily overlooked, but rather pretty, hairless perennial up to about 12cm tall. The basal leaves are twice ternate (with three lobes, each divided again into three lobes) and the stem leaves are ternate. The green flowers are unusual and distinctive, 6–8mm across, with four five-petalled flowers arranged in a square, each facing outwards – hence the alternate name of Town-hall Clock – and one four-petalled flower on top, facing upwards.

FLOWERING TIME April–May.

DISTRIBUTION Common in shady habitats throughout England, Wales and Scotland except the far north; absent from Ireland.

Moschatel *Adoxa moschatellina*

HONEYSUCKLE
Lonicera periclymenum

A strong-growing, deciduous, twining, woody climber, Honeysuckle reaches up to 6m tall, occasionally more with suitable support. The grey-green leaves are oval to elliptical, up to 7cm long, untoothed and growing in opposite pairs. The flowers are creamy yellow with purplish markings outside, becoming darker with age, narrowly trumpet-shaped, 4–5cm long, and have a protruding style and five stamens; they are clustered together into whorled terminal heads. Collectively, they are very attractive and fragrant, especially in the evening – a familiar and well-loved feature of the countryside in summer.

FLOWERING TIME June–September.

DISTRIBUTION Common and widespread throughout the area in woodland, hedgerows, roadsides, scrub and other habitats.

Honeysuckle *Lonicera periclymenum*

COMMON VALERIAN
Valeriana officinalis

An erect perennial up to about 1m tall, Common Valerian has stems bearing opposite pairs of large (up to 20cm long) pinnate leaves with narrow, toothed leaflets. The flowers are individually small (about 3mm across), pink or white and gathered together into large, terminal, umbel-like heads up to 12cm across. The fruits are feathery.

FLOWERING TIME June–August.

DISTRIBUTION
Common throughout the region in a variety of damp habitats including fens, meadows and damp woodland.

SIMILAR SPECIES
Marsh Valerian *V. dioica* is smaller, up to 30cm tall, with undivided (not pinnate) basal leaves. It is mainly southern, growing in calcareous wet places. **Red Valerian *Centranthus ruber*** has undivided, oval, greyish-green leaves, and red, white or pink flowers in dense terminal heads. It is an introduction from southern Europe, now widespread on rocks and walls.

Common Valerian *Valeriana officinalis*

WILD TEASEL
Dipsacus fullonum

Wild Teasel is an erect, robust biennial up to 2m tall, occasionally more, with prickly ridges on its stems. The basal leaves form a rosette covered with blister-based bristles; the stem leaves are large, ovate, joined around the stem in a cup and prickly on the midrib underneath. The flowers are small, pinkish-purple and gathered together into large, oblong or egg-shaped heads up to 8cm long, with a ring of long, thin, prickly bracts below them.

FLOWERING TIME July–August.

DISTRIBUTION Common on heavy soils in England and Wales and in open situations such as woodland clearings, stream banks and waste ground. It is rather uncommon in Scotland and Ireland.

SIMILAR SPECIES **Small Teasel *Dipsacus pilosus*** is smaller, reaching up to 1.5m at most, and the flower heads are white and more globular than the oblong heads of Wild Teasel.

Wild Teasel *Dipsacus fullonum*

FIELD SCABIOUS
Knautia arvensis

This erect, roughly hairy perennial is up to 1m tall. Its basal leaves are usually undivided, ovate and toothed; the stem leaves grow in opposite pairs and are pinnately divided, but there are often many slightly divided intermediate stages. The flowers are bluish-violet and small, with four petals and eight calyx teeth; they are aggregated together into large, flat heads about 3–4cm across with leafy bracts below them. The outermost flowers of the head are generally longer than the inner ones.

FLOWERING TIME July–September.

DISTRIBUTION Common and widespread throughout the area, though rare in northern Scotland and western Ireland, growing in grassy places on well-drained calcareous or neutral soils.

Field Scabious *Knautia arvensis*

SIMILAR SPECIES Small Scabious *Scabiosa columbaria* is rather similar, but smaller and more slender, with finer, more-divided leaves. The flowers have five petal lobes and five calyx teeth borne on smaller flower heads 2–3cm across. It grows in similar habitats, but is more restricted to calcareous soils, and rarely found outside England and Wales. **Devil's-bit Scabious *Succisa pratensis*** has smaller oval leaves, none of which is divided; the flowers are blue-purple, four-petalled, all the same size and borne in tight, rounded heads about 1.5–2.5cm across. It is widespread and common throughout the area in grassy places, damp or dry, on most soil types.

LESSER BURDOCK
Arctium minus

Lesser Burdock is an erect, leafy, branched biennial up to 1.2m tall. Its basal leaves are large, more long than wide, oval-triangular and up to 40cm long, with hollow leaf stalks; the stem leaves are alternate and smaller. The flowers are borne in short-stalked, oval to spherical heads, 1.5–2cm long (longer in fruit), consisting mainly of masses of narrow, overlapping, hooked bracts topped with a cluster of red-purple flowers. After flowering, these develop into the familiar sticky burs or stickyjacks.

FLOWERING TIME July–September.

DISTRIBUTION Widespread and common throughout the area in waste ground, roadsides, grassland and other habitats.

SIMILAR SPECIES **Greater Burdock A. *lappa*** has solid basal leaf stalks, and wider and flatter flower heads on longer stalks, in flat-topped inflorescences. It is less common and mainly southern.

Lesser Burdock *Arctium minus*

SPEAR THISTLE
Cirsium vulgare

This erect, spiny biennial is up to 1.5cm tall and branched in the upper part. The stems are downy-cottony with discontinuous spiny wings. The basal leaves are lance-shaped, deeply lobed and wavy-edged, with long, spine-tipped teeth; the stem leaves are smaller, with a marked long, pointed terminal lobe (the 'spear' of the name). The flowers are pinkish-purple and have spiny, oval heads up to 5cm long; they are solitary or borne in loose clusters. The seeds are feathery and known as 'thistledown'.

FLOWERING TIME July–October.

DISTRIBUTION Common everywhere in open, sunny habitats such as waste ground, roadsides, cultivated land and pastures.

Spear Thistle *Cirsium vulgare*

SIMILAR SPECIES Creeping Thistle *C. arvense* is a creeping perennial with erect, unwinged, spineless stems up to about 1m tall. Its leaves are lance-shaped, wavy and deeply lobed, with numerous spines on the margins. The flower heads are smaller, oval, up to 2cm long and topped with an attractive pom-pom of pinkish-purple flowers. This is a common and invasive weed in sunny habitats everywhere. **Marsh Thistle *C. palustre*** is similar in form, but has stems that are often purplish with continuous spiny wings. The flowers are darker reddish-purple (or occasionally white) with purplish-green bracts, borne in narrow heads up to 2cm long. It is widespread and common in damp, grassy places, but rarely invasive.

Creeping Thistle *Cirsium arvense*

COMMON KNAPWEED
Centaurea nigra

An erect, roughly hairy perennial up to 60cm tall, Common Knapweed is usually branched in the upper part. The leaves are narrowly oval, the lower ones being stalked and lobed or toothed, the upper ones simple and unstalked. The flower heads are rounded, 2–4cm across, solitary and terminal; they are made up of brown, fringed, membranaceous bracts topped with purple flowers that are normally all the same size.

FLOWERING TIME July–September.

DISTRIBUTION
Common and widespread in sunny, grassy habitats such as meadows and roadsides throughout Britain and Ireland.

SIMILAR SPECIES
Greater Knapweed
C. scabiosa is similar but often taller, with all leaves pinnately divided to a greater or lesser degree. The flower heads are similar, except that the outer florets are longer than the others, spreading out in a ring. It is locally common in calcareous grassland in England; uncommon elsewhere.

Common Knapweed *Centaurea nigra*

CHICORY
Cichorium intybus

Chicory is a perennial with erect, rigid, grooved, branching stems up to 1m tall. Its leaves are lanceolate, the basal ones stalked and often pinnately lobed or toothed, the upper ones simpler, clasping the stem at the base. The flower heads are bright blue, 2–4cm across, with all florets strap-shaped; they are borne in leafy, branching inflorescences. The flower heads open widely in the morning and fade during the day. The young leaves may be eaten as a salad.

FLOWERING TIME July–October.

DISTRIBUTION
Common in waste ground, meadows, roadsides, railroad sidings and other dry, grassy and open habitats, especially those with chalky soils, in England and Wales but decreasing in frequency northwards and westwards. The flower is not native to Britain.

Chicory *Cichorium intybus*

PERENNIAL SOW-THISTLE
Sonchus arvensis

This is a creeping stoloniferous perennial with erect, furrowed, branched stems up to 1.5m tall. The light-green leaves are all more or less pinnately lobed, alternate and edged with soft prickles, clasping the stem with rounded leaf bases. The flower heads are bright yellow, flat on top, about 4–5cm across and held in loose terminal clusters. The stems among the flowers, and all the bracts, are covered with sticky, yellow-tipped, glandular hairs. The stems and leaves discharge a bitter, milky fluid if they are broken.

FLOWERING TIME July–October.

DISTRIBUTION Very common and widespread throughout the region, except in higher mountain areas, in cultivated land, gardens (often as a weed), waste ground and coastal habitats.

SIMILAR SPECIES Dandelion *Taraxacum officinale* has similar flowers and seedheads, but it is borne on a leafless stalk and does not have the consistent ray florets of the Perennial Sow-thistle.

Perennial Sow-thistle *Sonchus arvensis*

GOAT'S-BEARD
Tragopogon pratensis

This is an erect biennial up to about 80cm tall, with unbranched or slightly branched stems. Its leaves are long and narrow, up to 30cm long, channelled, grass-like and long-pointed, with clasping bases. The flower heads are long-stalked, solitary and up to 5cm across, with bright yellow, rayed florets. The flowers open in the morning and close around midday (hence the plant's other name of Jack-go-to-bed-at-noon). Below each flower are about eight bracts, which in subspecies *minor* (the most common in the UK) protrude well beyond the florets; in subspecies *pratensis* (the most common in mainland Europe) they are about the same length.

FLOWERING TIME May–July.

DISTRIBUTION Common in grassy places in the lowlands, but rare in highland Scotland and much of Ireland.

Goat's-beard *Tragopogon pratensis*

SEA ASTER
Aster tripolium

Sea Aster is an erect, branched, hairless, slightly fleshy and short-lived perennial up to 90cm tall. Its leaves are narrow, linear, rather fleshy and up to 10cm long, half clasping the stem. The flowers are daisy-like and up to 2cm across, with bluish-purple rays surrounding a yellow disk; they are carried in an open, branched, rather flat-topped inflorescence.

FLOWERING TIME July–October.

DISTRIBUTION Always occurs in salty habitats, such as salt marshes, around the coasts of Britain and Ireland, rarely inland; common in suitable habitats.

SIMILAR SPECIES A number of naturalized Michaelmas daisies, such as **A. novi-belgii** are similar, but they are not fleshy and normally occur in non-saline habitats.

Sea Aster *Aster tripolium*

Common Fleabane
Pulicaria dysenterica

An erect, branching, softly hairy perennial arising from spreading runners and growing up to 60cm tall, Common Fleabane readily forms patches or clumps. The basal leaves wither by the flowering time; the stem leaves are alternate, ovate to oblong, usually untoothed and clasp the stem at the base. The flower heads are up to 3cm across, made up of a central circle of darker orange-yellow disk florets and an outer ring of golden-yellow ray florets; they are carried in loose, branching inflorescences. The plant was once used as a means of protecting houses from flea infestation; nowadays it is valued as an important nectar source for insects, and as an attractive garden plant.

FLOWERING TIME July–September.

DISTRIBUTION Common in England, Wales and Ireland in damp, open habitats such as riversides, fens, meadows and roadsides; rare or absent from most of Scotland.

Golden-samphire *Inula crithmoides*

Common Fleabane *Pulicaria dysenterica*

SIMILAR SPECIES Golden-samphire *Inula crithmoides* is a
somewhat similar, clump-forming plant, but has narrowly linear,
fleshy, toothed leaves; the flower heads are held in a loose, almost
flat-topped cluster. It is confined to coastal habitats, such as cliff-tops,
shingle and salt marshes, in England, Wales and Ireland, though it is
not common, except in a few favoured situations.

TANSY
Tanacetum vulgare

Tansy is a tall, erect, strong-smelling, hairless perennial up to 1m tall, which often forms dense patches. The leaves are alternate, feathery and pinnately lobed, with leaflets further lobed, and up to 20cm long. The flowers are yellow, with each head about 1cm across; they are gathered together into larger flat-topped inflorescences that are up to 15cm across. Tansy was formerly used as a medicinal and culinary herb, but it is now rarely used, as it can be poisonous if ingested in large quantities.

FLOWERING TIME July–October.

DISTRIBUTION Now naturalized or native in open, sunny habitats, such as roadsides, waste ground and grassland, throughout most of Britain, although it is uncommon both in northern Scotland and (not native) in Ireland.

Tansy *Tanacetum vulgare*

Oxeye Daisy
Leucanthemum vulgare

This erect, slightly branched plant is hairless or slightly hairy and up to 70cm tall. It has a basal rosette of oblong to spoon-shaped, stalked, lobed, toothed leaves; the stem leaves are alternate, unstalked, toothed and clasp the base of the stem. The flower heads are daisy-like, usually 3–5cm across, solitary and terminal, with a ring of white ray florets surrounding a yellow disc. It is sometimes known as Moon Daisy.

FLOWERING TIME May–September.

DISTRIBUTION Very common almost everywhere in the British Isles, except the highest mountains, in sunny, grassy habitats. Often planted as a component of seed mixtures.

SIMILAR SPECIES The garden plant **Shasta Daisy *L. x superbum*** is often naturalized; it is similar, but with larger flower heads, up to 10cm across, and unlobed lower leaves. A number of mayweeds and chamomiles are similar, but all are much more branched and shorter, and have slightly smaller flowers in branched inflorescences and more feathery, divided leaves. **Scentless Mayweed *Tripleurospermum inodorum*** is the most common species – an unscented, hairless, sprawling annual up to 50cm tall, with much-divided, feathery leaves. The flowers are similar, up to 4cm across, and borne in loose, branched inflorescences. It is abundant throughout the lowlands in disturbed habitats. A perennial, fleshier version is known as **Sea Mayweed *Tripleurospermum maritimum***, found in various coastal habitats.

Oxeye Daisy *Leucanthemum vulgare*

COMMON RAGWORT
Senecio jacobaea

This erect, hairless or hairy biennial or perennial is up to about 1m tall and usually branched in the upper parts. The basal rosette leaves are oval, pinnately lobed, rather crinkly and usually wither by the flowering time; the stem leaves are similar, clasping the stem. The flowers are bright yellow, with each head 1.5–2.5cm across, gathered into loose, flat-topped, branched clusters.

FLOWERING TIME June–October.

DISTRIBUTION Abundant and widespread throughout the region in sunny places, especially in abandoned or overgrazed pastures and waste ground.

SIMILAR SPECIES **Oxford Ragwort *Senecio squalidus*** is usually annual, more bushy, branched and up to 50cm tall. It is widely naturalized, from Italy, in open habitats throughout England and Wales, though rare in Scotland and Ireland.

Common Ragwort *Senecio jacobaea*

BUTTERBUR
Petasites hybridus

A low-growing, spreading perennial, Butterbur produces very large leaves after the flower heads open. The leaves are stalked, rounded to triangular with a heart-shaped base, distantly toothed and eventually enlarge to about 60–80cm across. The flowers appear before the leaves as erect, stiff, pinkish spikes, with scale leaves only, terminating in dense, cone-shaped inflorescences that are up to 30cm tall when in flower, and elongate considerably afterwards. The flower heads are reddish-pink, each up to about 1.2cm long. Male and female flowers are borne on separate plants, with the female flowers distinctly smaller (up to 6mm long). The large leaves were once used for wrapping butter – hence the English name.

FLOWERING TIME March–May.

Butterbur *Petasites hybridus*

DISTRIBUTION Common in damp habitats, such as riversides, wet meadows and alder woodland, throughout Britain and Ireland, though rare in northern Scotland. Male plants are far more widespread than female plants, which are only frequent in the north of England.

SIMILAR SPECIES Winter Heliotrope *Petasites fragrans* is an invasive introduced garden escape. Its leaves are smaller, up to 20cm wide and present at flowering time. The flowers are strongly scented and produced in late winter. It is widespread in roadsides and waste ground, usually near habitation.

COLTSFOOT
Tussilago farfara

A familiar and attractive early spring flower that can grow up to about 30cm tall, Coltsfoot produces distinctive bare flower spikes on purplish stems before the leaves appear. The leaves are stalked, triangular to heart-shaped (resembling horse's hooves, leading to the flower's name), toothed and up to 20cm across, appearing soon after flowering. The bright yellow flower heads are 2–3.5cm across, with numerous narrow rays; they are borne singly at the ends of short, erect, leafless stems up to 15cm tall. The leaves were formerly used medicinally, especially in the treatment of coughs and asthma.

FLOWERING TIME February–April.

DISTRIBUTION
Common and widespread in sunny, often disturbed habitats including waste ground, meadows, roadsides and riverbanks throughout the region. It tends to grow to greater heights in Scotland.

Coltsfoot *Tussilago farfara*

HEMP-AGRIMONY
Eupatorium cannabinum

This is an erect, strong-growing, downy perennial up to 1.2m tall. Its leaves are deeply lobed, palmate or trifoliate, with 3–5 narrowly elliptical to lanceolate, toothed lobes, basal and in opposite pairs on the reddish stems. The leaves resemble those of the hemp plant, hence its name, although the two species are not related. The flower heads are individually small, 3–5mm across, pink to purplish and grouped into large, rounded, branched inflorescences up to 15cm across. Collectively, these are very attractive to insects.

FLOWERING TIME July–September.

DISTRIBUTION
Common in a variety of damp or wet habitats such as fens, alder carr, riversides and lake margins throughout England, Wales and most of Ireland; rare in Scotland, where it is mainly western.

SIMILAR SPECIES
Could be confused with **Common Valerian** (*see* p. 136), but that species has pinnate leaves and separate, not composite, flowers.

Hemp-agrimony *Eupatorium cannabinum*

WATER-PLANTAIN
Alisma plantago-aquatica

Water-plantain is an erect, hairless, aquatic perennial up to 1m
tall. Its leaves are long-stalked, narrowly oval, pointed and up
to 20cm long; they are generally held erect above water level.
The flowers are small, up to 1cm across, with three white-to-
pink petals; they are borne in open, tiered whorls of branches,
each of which branches again. The flowers only begin to open
in the afternoon.

FLOWERING TIME August.

DISTRIBUTION
Common in damp
or freshwater sites,
especially where
there is periodically
exposed mud,
throughout lowland
areas of Britain
and Ireland; it is
less common in
Scotland, especially
in the north.

SIMILAR SPECIES
**Lesser Water-
plantain *Baldellia
ranunculoides*** is
smaller, up to 30cm
tall, with larger
flowers, usually borne
in a simple umbel, or
with one extra whorl
at most. It is scattered
throughout Britain
and Ireland, but
nowhere common.

Water Plantain *Alisma plantago-aquatica*

COMMON DUCKWEED
Lemna minor

The duckweeds as a group are tiny, floating, aquatic plants without clear divisions into stems and leaves; they are individually inconspicuous, but frequently form green aquatic 'lawns'. Common Duckweed has pale green, oval fronds that are flat on both surfaces and about 2–6mm long, each with a single root. The flowers are minute and rarely produced.

DISTRIBUTION Widespread and common in still and slow-moving waters throughout the region, except in mountain areas and the north of Scotland.

SIMILAR SPECIES **Ivy-leaved Duckweed *Lemna trisulca*** has narrow, slightly submerged, translucent fronds that are usually joined at right angles into loose colonies. It is common except in northern areas. **Greater Duckweed *Spirodela polyrhiza*** has larger fronds, up to 1cm long and reddish below, each with several roots. It is uncommon, and mainly southern.

Common Duckweed *Lemna minor*

LORDS-AND-LADIES
Arum maculatum

This is an unusual and distinctive, hairless, tuber-rooted perennial up to 40cm tall. The leaves are triangular, arrow-shaped, up to 20cm long, shiny, often spotted with black, crinkled and long-stalked, and appear in early spring. The flowers are unusual: they are grouped together in a narrow, cylindrical spadix, with the female flowers at the base, the male flowers above them, topped by a ring of sterile, bristle-like flowers, then terminating in a long, club-shaped, purplish (occasionally yellowish) appendix. The whole is enclosed by a large, greenish-yellow bract, the spathe, which wraps around the spadix, with the lower part closed off by the bristle-like sterile flowers. The flowers give off a meat-like smell that attracts flies; these push down through the ring of sterile flowers and pollinate the female flowers. They can eventually escape when the flowers wither, and may then carry pollen to another flower.

FLOWERING TIME April–May.

DISTRIBUTION Common in shady habitats in England, Wales and Ireland, though rare in Scotland.

SIMILAR SPECIES Italian Lords-and-Ladies A. *italicum* is similar, but has unwrinkled, clear green leaves, a clear yellow-green spathe that tends to flop over at the top and a shorter yellow spadix. It is uncommon, native in southern England only, but also occurring as a garden escape, especially as the white-veined form.

Lords-and-ladies *Arum maculatum*

159

Bog Asphodel
Narthecium ossifragum

This species has a very attractive flower, rather like a miniature lily. It is a creeping, hairless perennial with flower stems up to 35cm tall. The leaves are mostly basal, sword-shaped, flattened, curved, similar to iris leaves and reaching about a third of the height of the flowering stem. The stem leaves are few and small, sheathing the stem. The flowers are orange-yellow and 1–1.5cm across, with six 'petals' in a star shape; they are held in an erect inflorescence. The orange-red anthers have conspicuous orange-woolly filaments.

FLOWERING TIME July–September.

DISTRIBUTION A plant of peaty, neutral-to-acidic wet places, such as bogs and wet heaths, usually where there is slight water movement in the soil. It is abundant in places, but absent from large areas of eastern and central England.

Bog Asphodel *Narthecium ossifragum*

MEADOW SAFFRON
Colchicum autumnale

Meadow Saffron is a low-growing, cormous perennial that produces naked flowers in autumn and ripe fruits and leaves in spring. The flowers are wineglass-shaped, pink to purple and up to 20cm tall; they have six 'petals' and three styles, and a thin, whitish stem that is actually part of the flower. The leaves are shiny-green, oblong-lanceolate and up to 30cm long, carrying with them the fruit from the previous year. It is also known as Autumn Crocus, though true crocuses *Crocus* spp. have six styles and very narrow leaves; several species are naturalized.

FLOWERING TIME August–October.

DISTRIBUTION An uncommon plant of grassland and open woods, mainly in western England and eastern Wales, and very rare elsewhere.

Meadow Saffron *Colchicum autumnale*

SOLOMON'S-SEAL
Polygonatum multiflorum

This is an erect perennial with round, arching stems arising from a rhizome, reaching 80cm tall. The leaves are alternate, oval to elliptical, untoothed and unstalked, and clasp the stem. The flowers are tubular, constricted in the middle, greenish-white and 1–1.5cm long; they grow in pendulous clusters of 2–4 (occasionally more) in the leaf axils. The fruit is a blue-black, spherical berry.

FLOWERING TIME May–June.

DISTRIBUTION Locally common in dry woodland. Largely confined to England as a native plant, but widely naturalized elsewhere in the region.

SIMILAR SPECIES **Angular Solomon's-seal** *P. odoratum* is usually shorter and more erect, with angled stems and one to two longer, scented, not constricted flowers in each leaf axil. It is uncommon in rocky calcareous places in western England and eastern Wales.

Solomon's Seal *Polygonatum multiflorum*

FRITILLARY
Fritillaria meleagris

This attractive, erect, hairless herb has stems up to 40cm arising from an underground bulb. The leaves are narrowly linear, few, alternate and mainly up the stem. The flowers are distinctive and unmistakeable – pendulous, cup-shaped, 4–5cm long, usually solitary, and most commonly chequered pink and brownish-purple, but not infrequently all white. White flowers may amount to as much as 20% of the population in some places. It is also known as Snakeshead Fritillary.

FLOWERING TIME April–May.

DISTRIBUTION Uncommon, but locally abundant in damp floodplain meadows in south and east England, especially the Thames Valley. It is declining in the wild, but it is often grown in gardens and occasionally naturalized elsewhere.

Fritillary *Fritillaria meleagris*

RAMSONS
Allium ursinum

Ramsons is an erect perennial up to 45cm tall. The leaves are oval to elliptic, bright green, flat, and up to 20cm long and 7cm wide, growing on long stalks. The whole plant, especially the leaves, smells strongly of garlic when bruised. The flowers are white, starry and up to 2cm across; they are borne in dense, rounded terminal umbels, each with up to 20 flowers. It is also known as Wild Garlic and once used widely for culinary purposes, but is now used more rarely

FLOWERING TIME April–May.

DISTRIBUTION Common and widespread in most of the area – except in uplands or areas without woodland – in woods, along shady roadsides and in similar habitats.

Ramsons *Allium ursinum*

DAFFODIL
Narcissus pseudonarcissus ssp. *pseudonarcissus*

An erect, hairless perennial, the Daffodil often forms large clumps up to 35cm tall. The leaves are flat, narrowly linear, greyish-green, and about 1cm wide and up to 30cm long; they are held erect or arching. The distinctive flowers are solitary, yellow, terminal and turned sharply to one side. Each flower is made up of a ring of six pale yellow perianth segments (petals) with an inner deeper-yellow, trumpet-shaped corona 2–3.5cm long and about the same length as the petals. It is also known as Wild Daffodil and Lent Lily.

FLOWERING TIME March–April.

DISTRIBUTION Locally common in woods and old pastures, and on roadsides, in western England and eastern Wales as a native; rare elsewhere. It is occasionally abundant, for example in parts of Gloucestershire.

SIMILAR SPECIES Although daffodils are easy to identify as a group, they can be hard to identify as species, especially as there are many naturalized hybrids of garden origin. This is the only genuinely native species, best identified by its small flowers, short stature, difference in colour between its petals and corona, and its presence in more natural habitats.

Daffodil *Narcissus pseudonarcissus* ssp. *pseudonarmcissus*

LILY-OF-THE-VALLEY
Convallaria majalis

This low-growing, perennial plant has long, creeping rhizomes that give rise to erect pairs of oval-elliptical, parallel-veined leaves up to 20cm long. The flowers are borne in long-stalked, one-sided spikes up to 20cm tall; the flowers themselves are nodding, white, sweetly-scented and globose to bell-shaped, with six teeth. The fruits are small, red, spherical berries.

FLOWERING TIME April–May.

DISTRIBUTION An uncommon plant of dry calcareous habitats, especially ash woodlands on limestone, but also on limestone pavements and cliffs; occasionally in woods on sandy soil. It occurs mainly in northern Britain, but is completely absent from Ireland. It is also frequently grown in gardens, and not uncommonly naturalized.

Lily-of-the-valley *Convallaria majalis*

BLUEBELL
Hyacinthoides non-scripta

The Bluebell is one of Britain's most familiar and best-loved wild flowers. It is an erect, hairless, bulbous plant up to 50cm tall. Its leaves are narrow, strap-shaped, all basal, keeled below and hooded at the tip, and up to 30cm long. The flowers are blue (rarely white), narrowly bell-shaped and 1–2cm long, with tepals joined as a tube below, ending in six curled-back tips, and pendulous in few-flowered, drooping terminal inflorescences; the stamens have cream anthers. It is known as Wild Hyacinth in Scotland.

FLOWERING TIME April–May.

DISTRIBUTION Widespread and common in most of the area, normally in old woodland, but also grows in grassy places, especially near western coasts. It is usually abundant to dominant where it does occur.

SIMILAR SPECIES **Spanish Bluebell** *H. hispanica* is a common garden escape. It differs in having more erect flower spikes that are not one-sided, with more open, cup-shaped flowers without reflexed tips, and blue anthers (except in white forms). **Hybrid Bluebell** *H. x massartiana* (a hybrid between Bluebell and Spanish Bluebell) is another common garden escape, and also occurs naturally where the

Bluebell *Hyacinthoides non-scripta*

Spanish Bluebell *Hyacinthoides non-scripta x H. hispanica*

two species meet. It is intermediate in character between the above two species. As the hybrids are often fertile, a whole spectrum of variation occurs.

BUTCHER'S BROOM
Ruscus aculeatus

A curious and distinctive plant that is unlike any other native flower in the area, Butcher's Broom is a tough, erect, bushy, hairless, evergreen perennial that forms dense thickets up to about 1m tall. The 'leaves' on the stems are actually cladodes – flattened stems – that bear the flowers and then the fruit directly on them. The true leaves are reduced to tiny, inconspicuous scales. The cladodes are dark green, oval, leathery, spine-tipped and up to 2–3cm long, and they are borne alternately up the stem. The small, greenish-white, often purple-spotted, six-petalled flowers, each about 5mm across, are borne singly or in pairs in the centres of the cladodes. Male and female flowers are commonly, though not always, on separate plants. The female flowers develop into conspicuous, bright red, globose berries, about 1cm across. The stiff, spiny branches were once used for cleaning off butcher's boards – hence the name – and are sometimes used as Christmas decorations.

Butcher's Broom *Ruscus aculeatus*

FLOWERING TIME January–April, with berries usually
present in autumn and winter.

DISTRIBUTION Native only in southern England and Wales, in
dry woods, hedgerows and non-saline coastal habitats, but also
naturalized widely elsewhere; they are very rare in Ireland.

BLACK BRYONY
Tamus communis

A climbing, hairless, herbaceous perennial, this species twines clockwise up to heights of 5m or more. The leaves are heart-shaped, bright shiny green, net-veined and long-stalked; neither the leaves nor the stems have tendrils. The flowers are small, green, starry and six-parted; they are borne in short racemes from the leaf-axils. Male and female flowers are produced on separate plants (known as dioecious). The male flowers are stalked, the female flowers unstalked. The berries are conspicuous bright shiny red, almost spherical, about 1cm long and poisonous. This is the only north European member of the otherwise mainly tropical yam family.

> **FLOWERING TIME** May–July, with berries visible through the autumn into early winter.

DISTRIBUTION Common throughout England and Wales in open woods, scrub, hedgerows and similar habitats, but virtually absent from Scotland and Ireland.

Black Bryony *Tamus communis*

White Bryony *Bryonia dioica*

SIMILAR SPECIES White Bryony *Bryonia dioica* is a rather similar but unrelated twining plant, also dioecious. It has softer, paler green, palmately lobed leaves, with long, spirally coiled tendrils arising from the bases of the leaf stalks. The flowers are also pale whitish-green, but have five petals. Habitats and distribution are similar, though the plant is more confined to calcareous soils.

Yellow Iris
Iris pseudacorus

This is an unmistakable erect perennial, with tall stems up to 1.5m in height arising from strong-growing, creeping rhizomes. The leaves are long, narrow, sword-shaped and grey-green, with a distinct midrib. The flowers are bright yellow and large, up to 10cm across, with three broad, oval falls (the equivalent of sepals) veined with purple, and three shorter obliquely erect standards. It is also known as Yellow Flag.

FLOWERING TIME May–July.

DISTRIBUTION Widespread and generally common throughout the region in most wet habitats, including wet woodland, riverbanks, meadows and dune-slacks.

SIMILAR SPECIES Stinking Iris *Iris foetidissima* is smaller, with dull purplish-yellow flowers, and all parts smelling of roast beef when crushed, giving it the alternate name of Roast Beef Iris. It occurs in dry habitats mainly in southern areas of the region.

Yellow Iris *Iris pseudacorus*

Common Twayblade
Neottia ovata (formerly *Listera ovata*)

Common Twayblade is an erect, downy perennial up to 60cm tall, with a distinctive form: the single green stem bears a pair of broadly oval, unstalked, ribbed, horizontal or ascending leaves, which are held a few centimetres above the ground but below the middle of the stem. These leaves give the plant its name, 'twayblade', meaning 'two leaves'. The stem is hairy above and hairless below these leaves. The green flowers are borne in a thin, cylindrical terminal inflorescence. Each flower has five roughly equal perianth segments forming a hood, and a longer sixth one that hangs down in the form of a forked lip.

FLOWERING TIME June–July.

DISTRIBUTION
One of the most common orchids, found in meadows, woods and roadsides, usually on neutral-to-calcareous soil, throughout all of Britain and Ireland, although it is less common in the north of Scotland.

Common Twayblade *Neottia ovata*

LADY'S-SLIPPER
Cypripedium calceolus

This strikingly attractive and distinctive plant can hardly be confused with any other wild flower in the region, if you are lucky enough to find it. It is an erect, downy and perennial herb up to 60cm high when in flower. The leaves alternate, oval to lanceolate, are up to 18cm long, clasping the stem, and noticeably pleated and veined. The flowers are borne singly or in pairs, usually terminal, with a large leaf-like bract just below the flower. Each flower is very large, up to 10cm across the longest dimension. There is a large maroon-brown sepal at the top of the flower and two fused sepals below it. The upper two petals are similar in colour to the sepals, though narrower. The lower petal is in the form of a large yellow lip – the slipper of the name – with bulging sides and an opening at the top. Unusually among European orchids, the flowers attract insects – mainly solitary bees of the genus *Andrena* – and initially trap them by the structure of the interior of the lip. Eventually, the bees escape through one of two rear exits, but only after they have rubbed themselves against the stigma, and collected new pollen to take elsewhere.

FLOWERING TIME May–June.

DISTRIBUTION A rare plant of upland limestone areas, especially in light woodland. Confined, in Britain, to the Yorkshire Dales, but more widespread further north in Europe, and in limestone mountains further south.

Lady's-slipper *Cypripedium calceolous*

180

BEE ORCHID
Ophrys apifera

The distinctive flowers of the bee orchid group *Ophrys* spp. are highly evolved to attract potential pollinators. The Bee Orchid's erect perennials are up to 40cm tall, with greyish-green leaves in a persistent basal rosette and a few leaves decreasing in size up the stem. The large flowers are borne in loose spikes of up to 10 (more normally 3–7). The three, outer petal-like sepals are rosy pink or whitish, pointed and up to 1.5cm long; the two upper petals are short and hairy, less than half the length of the sepals, with the lower petal, the labellum, in the form of a broad, rounded, concave, furry, brown, bee-like structure about 1–1.5cm long and up to 1cm wide. The general impression is of a bee on a flower. The flowers have evolved to attract specific male Hymenoptera that attempt to mate with the flower, thus carrying out pollination. In this species, the mechanism usually fails and the flowers are regularly self-pollinated.

FLOWERING TIME June–July.

DISTRIBUTION Rather common throughout England in dry, open, sunny, calcareous habitats such as chalk downland, old quarries, dunes and roadsides. They are uncommon in Wales and Ireland, and virtually absent from Scotland.

Bee Orchid *Ophrys apifera*

EARLY PURPLE ORCHID
Orchis mascula

An erect, hairless perennial, this orchid dies back completely after flowering. The flowering stems are robust and often slightly purplish. The leaves are shiny, narrowly oblong and dark green, often with longitudinally aligned purplish blotches (though they may also be unspotted); they are clustered towards the base of the stem. The flowers are borne in conical-triangular, dense terminal heads, and open from the bottom first. The flowers are red-purple with two erect outer sepals, the remaining sepal and two petals forming a hood, the third petal a broad, pendulous three-lobed lip, often with a paler central patch dotted with blackish-purple spots. Behind the flower there is a stout, upwards-curving, nectar-filled spur.

FLOWERING TIME April–June.

DISTRIBUTION Widespread in a variety of neutral to calcareous habitats, including woodland, meadows and roadsides, throughout most of Britain and Ireland.

Early Purple Orchid *Orchis mascula*

Green-winged Orchid *Anacamptis morio*

SIMILAR SPECIES Green-winged Orchid *Anacamptis morio* is similar (it was formerly classified as an *Orchis* species) except that it is usually shorter, up to 30cm tall, and always has unspotted leaves. The flowers are similar, but all three sepals form a tight helmet and they are striped inside with green veins and (sometimes) pale stripes. A mainly southern species, it is widespread in grassy areas in England, Wales and Ireland, but virtually absent from Scotland and Ulster.

PYRAMIDAL ORCHID
Anacamptis pyramidalis

This erect, hairless plant has unbranched, single stems up to 30cm high. The leaves are lanceolate, narrow, pointed and clustered as a rosette in the lower part of the stem, with a few smaller ones higher up. The flowers are pink and borne in a dense, conical to triangular spike that elongates as it ages. They are small and unspotted, with the upper sepal and two upper petals forming a helmet, while the third petal, the lip or labellum, is larger (to 6mm), hangs downwards and has two erect ridges on it. Each flower has a long, slender spur up to 1.2cm in length.

FLOWERING TIME June–August.

DISTRIBUTION Common in England and Ireland, but scarcer in Wales and Scotland, in dry, sunny, grassy, calcareous habitats such as chalk downland and stabilized sand dunes.

SIMILAR SPECIES Fragrant Orchid *Gymnadenia conopsea* is rather similar in having pink, long-spurred, unspotted flowers, but differs in having a longer cylindrical flower spike, and a curved and even longer spur up to 2cm long. There are no raised ridges on the labellum. The flowers are fragrant, though individual plants vary. The species is widespread in similar habitats, though it has a greater tolerance of dampness and differing soil acidity.

Pyramidal Orchid
Anacamptis pyramidalis

Fragrant Orchid
Gymnadenia conopsea

COMMON SPOTTED-ORCHID
Dactylorhiza fuchsii

An erect, slender, hairless perennial, this species is up to 40cm tall. The leaves are greyish-green, spotted with transversely aligned dark blotches (see also Early Purple Orchid, p. 182). The lowest leaves are oblong-lanceolate and blunt, and the upper leaves narrower and more pointed. The flowers are pale pink to pale purple (occasionally white), with looped darker dashes and dots. The outer petals are spreading, the upper sepal and upper two petals forming a loose hood. The largest petal, the labellum, is about 1cm long and deeply divided into three lobes, of which the central one is the longest; the spur, projecting behind the plant, is narrowly conical and up to 1cm long. The flowers are held in a long, dense, cylindrical spike.

FLOWERING TIME June–August.

DISTRIBUTION Widespread and common almost throughout the area in a variety of neutral to base-rich habitats such as chalk downland, open woods, meadows, fens and roadsides.

SIMILAR SPECIES Heath Spotted-orchid *D. maculata* is similar, but has narrower leaves with rounder spots, and flowers in generally shorter, wider inflorescences. The labellum is much less divided, with a short, small central lobe and two larger, rounded side lobes. It is widespread in damper, more acid habitats such as wet heaths and moorland. These spotted orchids can be quickly differentiated from the preceding orchid species by the leaf-like bracts in the flower spikes.

Common Spotted-orchid *Dactylorhiza fuchsii*

Glossary

acid, base-rich: describes the acidity of a soil. In general, acid soils develop over rocks such as granite, schist or sandstones, while base-rich soils develop on limestones, chalk or dolomite. This greatly affects what plants can grow there.

axil: point where a leaf-stalk joins the main stem.

bract: small, scale-like or leaf-like structure that supports a flower.

bracteole: bracts supporting an upper ring of flowers in the inflorescences of the carrot family (*see* pp. 100–101).

bulbil: small, bulb-like structure, often produced in place of flowers, that can drop off and grow directly into a new plant.

calcareous soil: soil that has a high content of lime, as on chalk.

carr: woodland on habitually wet soil, such as alder woodland around a lake.

composite: the flowers of the Asteraceae family, which are made up of tight clusters that collectively look like a single flower, such as Sea Aster (*see* p. 147).

disk and ray florets: the different types of flower found in composite flowers (*see* p. 151). For example, in a daisy, the central small yellow flowers are disk florets, while the ring of white 'petals' are ray florets.

follicle: dry fruit formed from one carpel, which splits along one side to release the seeds. It is typical of paeonies and some members of the buttercup family.

Hymenoptera: insects from the hymenopteran order, such as bees, wasps and sawflies.

inflorescence: distinct cluster of flowers, such as the spike of an orchid or the umbel of a hogweed.

naturalized plant: species of plant that is not native to a region, but that now self-seeds and spreads widely.

nectaries: small structures, often located at the base of a petal, which produce nectar.

palmate: leaf shape that has several lobes (usually 5–7) with veins that radiate from a central midrib, like fingers from an outstretched palm (*see* photo above).

petal: *see* **Introduction**, p. 8.

pinnate: leaf shape with leaflets on either side of the stem, usually in pairs opposite each other, resembling a feather or plume (*see* photo below).

raceme, spike: narrow, unbranched inflorescences with the flowers opening from the bottom upwards; spikes have unstalked flowers.

sepal, calyx: *see* **Introduction**, p. 8.

spur: outgrowth of a petal or sepal in the form of a tube closed at one end.

stamen, anther, filament: *see* **Introduction**, p. 8.

stipule, style: *see* **Introduction**, p. 9.

trifoliate: leaf divided into three parts, for example clover leaf.

umbel: inflorescence of short flower stalks, similar in length, spreading from a common centre (*see* photo right).

Bibliography and Resources

There are many books dealing with various aspects of European wild flowers. Those listed below will lead you further into this extensive literature.

Blamey, M. and C. Grey-Wilson, *Cassell's Wild Flowers of Britain and Northern Europe*, Cassell, 2003. A large-format, fully illustrated guide to most of the flowers of northern Europe.

Gibbons, B., *Wildflower Wonders of the World*, New Holland, 2011. Gives details of some of the best places to see flowers throughout the world.

Rose, F. and C. O'Reilly, *The Wild Flower Key*, Warne, 2006. A well-illustrated portable guide to the flowers of Britain and nearby parts of Europe, with many detailed keys.

Stace, C., *New Flora of the British Isles*, third edition. Cambridge University Press, 2010. The most complete and up-to-date flora of the British Isles.

Organizations and Societies

Below are the names and websites of various organizations and societies that promote an interest in wild flowers, with a brief note on the particular interests of each society.

Botanical Society of the British Isles (BSBI)
www.bsbi.org.uk
The main society for the study and recording of wild plants in Britain and Ireland.

Field Studies Council (FSC)
www.field-studies-council.org
An educational charity that runs courses including many on plant identification at a number of field centres around the UK.

Wildlife Trust
www.wildlifetrusts.org (for UK);
http://iwt.ie/ (for Ireland)
The county wildlife trusts are conservation organizations working towards the protection of species and habitats in every UK county and the whole of Ireland.

Plantlife
www.plantlife.org.uk
The major UK and European charity dedicated to the protection of wild plants.

Index